AF697965

"All Nature Is But Art"

LOCUST HILL LITERARY STUDIES
NO. 12

"All Nature Is But Art"

The Coincidence of Opposites in English Romantic Literature

Mark Trevor Smith

LOCUST HILL PRESS
West Cornwall, CT
1993

© 1993 Mark Trevor Smith
All rights reserved

PR
457
.S59
1993

Library of Congress Cataloging-in-Publication Data

Smith, Mark Trevor.
All nature is but art : the coincidence of opposites in English romantic literature / Mark Trevor Smith.
282p. cm. -- (Locust Hill literary studies ; no. 12)
Includes bibliographical references and index.
ISBN 0-933951-44-2 : $32.00
1. English literature--19th century--History and criticism. 2. Romanticism--Great Britain. 3. English literature--18th century--History and criticism. 4. Blake, William, 1757-1827--Aesthetics. 5. Opposition, Theory of, in literature. 6. Coincidence in literature. 7. Paradoxes in literature. 8. Polarity in literature. 9. Oxymoron. I. Title. II. Series.
PR 457.S59 1993
820.9'145--dc20 93-27166
CIP

Printed on acid-free, 250-year-life paper
Manufactured in the United States of America

Contents

Acknowledgments

David Erdman has caused or helped most of the happy events in my career. A brief anecdote will reveal his easy graciousness. When I arrived at Stony Brook in 1977, my first stop was to visit Erdman, for whose inspiration and instruction I had signed up. A few days into the semester, I was walking in the hall past a classroom in which he was teaching. At the end of the hall stood a graduate student named David. Not even thinking about Dr. Erdman's first name, I called out, "Hi, David." The next time I walked past that class, David Erdman called out cheerily, "Hi, Mark." I blushed when I realized what was happening: he thought that I had interrupted him in the middle of class to yell out his first name. I then blushed again in honor of David's insult-proof egalitarianism. Multiply that graciousness by thousands of his students and colleagues, add his massive scholarship, and you see a man who deserves even more honors than he has already received.

During the summer of 1991 I studied with Leo Damrosch in his NEH seminar devoted to Rousseau and Blake. His *Symbol and Truth in Blake's Myth* had already prodded me more than any other book to try to formulate my own ideas. During the seminar, he handled my disagreements with good humor. His self-described eclecticism and marginality helped make him a perceptive interpreter of Blake (and of Rousseau in that seminar). This book is in large measure an attempt to answer not only Damrosch's book and its strong philosophical basis, not only Damrosch's refusal to grant Blake's mystical vision, but also his persuasive voice, skeptical eyes, and bemused laugh.

During the summer of 1985 I studied with Michael Cooke in his NEH seminar, "English Romanticism: The Problem of Wholeness." His *Acts of Inclusion* had already inspired me with its similarities to my own ideas. During that seminar, I completed an essay, "Striving with Blake's Systems," which appeared in *Blake and His Bibles*, edited by David Erdman, published by Locust Hill Press. That essay was really the seed of this book and could easily be a chapter of it. Michael was killed in a car accident on September 14, 1990. Although I do not converse with him daily, as William did with Robert, I often hear his encouraging voice, see his beaming smile, and feel his religious laughter.

I could dedicate this book to any one of those three, but I do dedicate it to Sonja Lallemand, with whom I have been joined in a passionate *coincidentia oppositorum* for over a decade. My daughters, Priscilla Barbara Smith—who now knows more about Buddhism than I ever will—and Catherine Blake Smith—who loves to recite the Introduction to *Songs of Innocence*—make every day glad. From the beginning, my parents, Rex and Barbara Smith, made books as essential as food.

Other people have helped me in large ways and small, from substantial advice to chance remarks that gave me renewed strength to accomplish this task: Pat Skarda and the other members of NEH summer seminars in 1985 and 1991, Ron Lunsford, Katherine Lederer, Don Holliday, Jim Jones, Tita Baumlin, Dean Hinnen, Bill Burling, Anthony Vital, Ruth Goehring, Sharon Dingman, Judy Reynolds, Thomas Maresca (who first introduced me to Nicholas of Cusa as a way of reading Pope), Thomas J.J. Altizer, Lee Miller, Betty T. Bennett, David Sheehan, Jerome Christensen, George Gilpin, Pam Senter, Maria Cecilia Fortou, Michael Tolley, Eric Link, Phill Pulsiano, Paula Uruburu, Lyn Yonack, Duane Bedell, Ellen Gardiner, Janice Queen, Karen Otis, Duane Coltharp and many other students in my classes, and the librarians at Harvard, Stanford, and Southwest Missouri State University.

Tom Bechtle, publisher at Locust Hill Press, has shown me the most extraordinary patience. Let his greatest reward be the knowledge that my patience toward others has increased by his example.

Jim Baumlin, not for the first time, performed beyond the call of friendship.

The two summers that I spent studying in NEH seminars stimulated my intellect more than any other experiences in my life. I thank the National Endowment for the Humanities for their wise and generous support of their summer seminars for college teachers.

The final two chapters of this book have been re-worked from my dissertation, "William Blake's Transfiguration of the Bible in *Jerusalem*." The chapter on Pope has been re-worked from an essay published in the journal of the Midwestern Modern Language Association; the chapter on Shelley has been re-worked from an essay published in the journal of the Missouri Philological Association; the chapter on Mary Shelley has been re-worked from an essay published in the journal of the Arkansas Philological Association.

Introduction
The Tool of the Imagination

If the Imagination can stand as the best concept with which to define Romanticism, then the pursuit of the coincidence of opposites, one of the favorite tools of the Imagination—if not indeed its only tool—can stand as a fundamental activity of writers of the Romantic era. The power of the coincidence of opposites to generate the visions of Romanticism arises from its simplicity and its profundity. What could be simpler than to say "black" in response to "white"? What could be more profound than to say that God becomes man or Jesus becomes Satan?

The urging toward coincidences of opposites in the writers that I discuss here originates from a discontent with the way things are. Nicolas Cusanus, Alexander Pope, Samuel Taylor Coleridge, Mary Shelley, Percy Bysshe Shelley, and William Blake all lament ordinary, accepted ways of living and perceiving; they all insist that we should change our ontology and therefore our epistemology by acknowledging that logical opposites, which by standard Aristotelian thinking are mutually exclusive, can indeed coincide.

To the twentieth-century ear, the word "coincidence" smacks of chance and randomness. To call something a coincidence implies that it should not really happen and thus allows us to dismiss the coincidence as an unexpected, unusual anomoly that, according to probability, will not happen again. The history of the word in English, as chronicled in the *Oxford English Dictionary*, reveals that this dismissive meaning gradually creeps in, as the word is linked with "casual" and "undesigned." The root meaning of the Latin word, which was used in the seventeenth century as a verb in English in its Latin form, "coincidere," is simply "to occur together."[1] It was a term in astrology. When two planets come together in the sky in a conjunction, they coincide. For example, Venus and Mars, whose basic energies are opposed, can come together in conjunction. Neither one necessarily over-rules the other, but neither are they reconciled. The two opposing forces, occupying the same space at the same time, interact in dynamic interchange.[2] They exert their opposing forces in a coincidence of opposites. From an astrological perspective, such a coincidence is not an improbable accident of chance, but an essential element of God's total artistic pattern of the universe. As we lost astrology from our universe, we also lost the most basic meaning of coincidence. "Having no apparent causal connexion" gradually became the most common meaning of the word. So to the modern lack of imagination the conjunction of Venus and Mars is a meaningless coincidence. In fact, from that perspective, the two planets do not really occupy the same space at the same time; they simply appear close from our vantage point.

When the writers in this study explore the *coincidentia oppositorum,* they are not only working to bring opposites together, but they are working from an assumption that coincidence is not chance, but pattern. Pope, without using the word "coincidence," virtually says as much. Thus, the struggle of these writers is twice as hard as one might expect: not only are they working against the powerful Aristotelian tradition that logically excludes the coincidence of opposites, but they are also working against a trend in which the very word which basically means "occurring together because these things are part of a total harmonious plan" is coming more and more to mean "occurring by unimportant chance because in this unfathomable universe you might be unlucky enough to get struck by lightning."

The most frequent emphasis in the definitions of "coincide" and "coincidence" in the OED is on the occupation of space:

> Coincidence ... the occupation of the same place or part of space.
>
> Coincide ... To fall together and agree in position; to occupy the same area or portion of space (as e.g. the superposed triangles in Euclid I.8); to be identical in area and position. Said of points, lines, or any geometrical magnitudes.

As we shall see, Nicolas Cusanus works especially from principles of Euclidean congruence to develop his divine geometry, in which the infinite line and the infinite circle coincide. For now we can note that if two finite triangles coincide so perfectly as to be exactly superimposable, they will appear to be one. From a Newtonian perspective, such oneness would be only an optical illusion, much like the illusion from astrology when Venus and Mars appear to coincide. When Isaac Newton describes his experiments, as quoted in the OED, he implies just such a perspective:

> 1704 Newton Opticks 1 J These circles .. as I went from them, they came nearer and nearer together, and at length became coincident.

In the universe of Newton, such a coincidence is merely an appearance. Obviously the circles did not really become one; they just looked like one from the observer's skewed perspective. The writers here will also play with such perspectives, especially in mirror images, but beyond the games on the surface, these writers are searching for a bedrock of coincidences of opposites in which the oneness does not arise like some optical illusion just because of the relative values of space and time, but because of the fundamental structures of the universe, both physical and spiritual, beyond space and time.

In Chapter 1, I set out some of the key concepts of the fifteenth-century monk who is credited with developing the concept of the *coincidentia oppositorum* in his *De Docta Ignorantia* (*Of Learned Ignorance*). Essentially Nicolas Cusanus argues that finite logic prohibits us from understanding much about God, Who by His very perfect completeness must include coincidences of opposites. Although Cusanus seems content to leave most of the ways of God unfathomable to man, he does point toward more radical joinings of man and God that Blake will fulfill. In Chapter 2, I show how Alexander Pope's *Essay on Man* sets man in relation to a universe, the apparent disorder of which is really God's artistic plan. Like Cusanus, Pope insists that only by confounding ordinary logical restrictions can man begin to accept the totality that includes him. When I originally decided to include Pope, I thought that he would provide a useful neo-classical, conservative contrast to the later Romantics. Although he does provide such a contrast, I have discovered that many of his pronouncements are remarkably similar not only to Coleridge's but even to Blake's. In Chapter 3, I discuss Samuel Taylor Coleridge's explicit search for the reconciliation of oppositions in his prose and poetry, and I conclude with a late poem dedicated to the eleventh-century monk Berengarius, who, in a way that Coleridge is particularly sympathetic to, tried but failed to maintain his vision of one particular coincidence of opposites, the Real Presence in the Eucharist. Those first three writers all maintain, to varying degrees and with various emphases, a fundamentally conservative view of God's universe. Although Coleridge, as the first Romantic writer in my study, tries to break through to a greater participation in the spiritual world, he finally acquiesces in mystery, as do Cusanus and Pope, as the final condition of man and universe.

Chapter 4, on Mary Shelley's *Frankenstein*, comes in the middle, halfway between the conservatism of Cusanus, Pope, and Coleridge, and the radicalism of Percy Bysshe Shelley and William Blake. With attention especially to the concentric circles of the narrative technique, I explore the opposition between the selfish egotism that blatantly strikes every reader and the less obvious selflessness that goes beyond even friendship, especially in the critically neglected tale of the noble fellow near the beginning of the novel. Although *Frankenstein* does present basically a pessimistic view of the human condition, its countertheme of disinterested benevolence constructs a dynamic attempt to find a coincidence of opposites. In Chapter 5, the opposition between self and other emerges in several poems by Percy Bysshe Shelley. After the disasters of solipsism in *Alastor* and the triumph of coincidences of opposites in *Prometheus Unbound*, I conclude with the unresolved struggle between outer and inner in *The Triumph of Life*.

Chapters 6, 7, and 8 are devoted to William Blake. In Chapter 6, I analyze "The Mental Traveller," an archetypal presentation of the absolute failure of the oppositions between male and female to come to any resolution as they spin around in their circles of mutual hatred. In Chapter 7, I explore Blake's internal eternity, especially in *Jerusalem*, and the methods of self becoming other that Blake urges to help us perceive and live that internal eternity. In Chapter 8, I focus on "Monos o Iesous," the Greek inscription on plate 4 of *Jerusalem*, to show how the simultaneous transcendence and immanence contained in that phrase lead to a totally inclusive vision of the coincidence of opposites.

Clearly my approach is not chronological, but instead situates these writers along a continuum of how far they go in exploiting the revolutionary potential of the *coincidentia oppositorum*. Although the order of my presentation thus contains a certain logic, each chapter is fundamentally self-contained and therefore can be read on its own. The reader who skips over the chapters on Mary Shelley and Percy Bysshe Shelley may find the thread of the argument clearer, but she will miss some marvelous complications.

Through all eight chapters I point out similarities in the writers' confrontations with the question of the *coincidentia oppositorum*, especially in the ways that they use the imagery of circles to give us geometrical pictures of abstract concepts. In each writer, those circles arise from the interaction of two basic universal forces: a centripetal, implosive force that works toward the center and promotes self-preservation, and a centrifugal, explosive force that pushes out the circumference and urges self-annihilation. Since the former force is the one that "naturally" predominates in humans, these writers all emphasize the latter, not in order to replace one force with the other, but to find a fruitful interaction between them.

Although the principle of the *coincidentia oppositorum* that I explore here has many similarites with other binary systems, I must distinguish it from some of the most obvious other systems that might spring to mind. First, although Hegelian patterns can be useful by analogy, finally the *coincidentia oppositorum* does not search for a *tertium quid* to resolve the tension. Reconciliation, although sometimes the goal of some of these writers, is not essential, and in fact is explicitly eschewed by Blake: "Whoever tries to reconcile them [opposing forces of Devouring and Prolific] seeks to destroy existence."[3] The word "oppositorum" in *coincidentia oppositorum* requires a dynamic back-and-forth, springing from irreducible alterity. At the other extreme of binary criticism, deconstruction can provide worthwhile analogies:

> Far from being a chain which moves deeper and deeper into the text, closer and closer to a definitive interpretation of it, the mode of criticism sometimes now called 'deconstruc-

> tion,' which is analytic criticism as such, encounters always, if it is carried far enough, some mode of oscillation. In this oscillation two genuine insights into literature in general and into a given text in particular inhibit, subvert, and undercut one another. This inhibition makes it impossible for either insight to function as a firm resting place, the end point of analysis.[4]

Although the dynamic character of deconstructive moves is similar to that of the *coincidentia oppositorum*, the *coincidentia oppositorum* does not revel in endless oscillation. Instead the word "coincidentia" requires a complete coming together, a copulative interpenetration.

In this combination of coincidence, which always unifies, and opposites, which always repel each other, lies the fundamental paradox, with which the finite mind can stretch itself to a larger, more comprehensive view of existence. But that stretching is not simply a shift in perspective, by which to alter epistemology. Although each of these writers focusses intently on issues of sight, and often on issues of light that aids or baffles that sight, the epistemology that they urge is a by-product of a new ontology. For if one's being in the universe changes, then one's perspective changes also. And vice versa. Blake brings to a climax a view of the universe that these other writers approach: accepting the coincidence of opposites gives us a whole new universe to see and a whole new ontology of living in that universe. In the thematically opposed stories of the transfiguration and of the woman taken in adultery, which I discuss in Chapter 8, Blake presents a revolutionary combination of transcendence and immanence. The opposites come together in a way that can propel us into a new world, which, as at the end of *Prometheus Unbound* or the end of *Jerusalem*, is exactly the same world as before, yet at the same time totally different. Sameness in difference, and difference in sameness.

I do not attempt to escape what Jerome McGann describes as the Romantic Ideology:

> scholarship and criticism of Romanticism and its works are dominated by a Romantic Ideology, by an uncritical absorption in Romanticism's own self-representations.... Romantic

> poetry incorporates Romantic Ideology as a drama of the contradictions which are inherent to that ideology.[5]

Still less do I escape what Anne Mellor calls "masculine" Romanticism:

> a binary model is already deeply implicated in "masculine" Romanticism.... The principle of polarity ... requires the construction of an Other which is seen as a threat to the originating subject.... the women writers of the Romantic period resisted this model of oppositional polarity ... for one based on sympathy and likeness.[6]

Mellor herself questions her own method of employing a binary structure, which she says is necessary on the way toward learning to "think beyond a dialectic based on polarities."[7] In fact, I contend that the coincidence of opposites, especially in the hands of Blake, incorporates not only struggle against an Other in a masculine model of oppositional polarity but also, and just as strongly, a feminine model of sympathy and likeness with that Other. While I agree that we need to think beyond polarities, or at least to try to, the practice of thinking into and with polarities is one method that will help us think beyond. "If the fool would persist in his folly, he would become wise." This book is my folly; perhaps one day I will achieve a prospect from which, as Jerome McGann has done, I can repudiate my former errors. Meanwhile, I plunge in, willing to fall rather than not make definite shapes and minute particulars.

The imagination as I explore it through the coincidence of opposites refuses to be bound by rationality—it does not seek a middle ground between extremes. It includes extremes. It embraces paradox. The simplest and most fundamental way of rising from ordinary perception to imaginative perception is by seeing through the false limits imposed by an abstract logic that denies that A and not-A can exist at the same time and to the same extent.

The logic of the human mind and its attempts to find or establish institutions and other forms of order, need to work in dualisms—us/them, inside/outside, good/evil, etc. Such binarism is widely condemned by many literary critics today, as they seek new positions that are more inclusive. While emphases on inclusion and diversity take us in the right direction, they can become static and even dogmatic. The implications of the coincidence of opposites lead toward dynamic open-mindedness. The coincidence of opposites requires a radical combination of otherness and selfhood. The imagination must hold—without clutching—the unity of all humanity in one—called Albion by Blake—and at the same time the absolute individuality of each person and therefore the irreducible otherness of individuals and the world. This paradoxical combination opens a world of inclusiveness, tolerance, and creativity.

Of course all theories of literature, even the most inclusive ones, remain too small. The coincidence of opposites, however, provides a dynamic framework for thought that refuses to accept its own boundaries. I do not intend in my emphasis on the coincidence of opposites to assert closure or any kind of dogmatic system or "key" to Romanticism or to Blake. But I do assert that the coincidence of opposites is an often neglected tool for opening up Blake and Romanticism.

The coincidence of opposites serves as a fundamental and inclusive theory to incorporate and confront not only older methods of thought but also to acknowledge and respond to recent trends such as dialogism and deconstruction. Like deconstruction, the coincidence of opposites destroys complacency with the simplest of techniques: slap-in-the-face denial:

> Blake's technique of contrary or diabolical reading … which proceeds by inverting the privileged oppositions in a text (as in his transvaluation of Milton's devils and angels) and then by calling into question the whole system that keeps them opposed, seems to me very close to the method of deconstruction.[8]

And like dialogism, the coincidence of opposites constructs systems in the most comprehensive way by allowing various voices to be heard. Refusing to tolerate the soft edges of compromise and smiley faces, it produces a hard edge that escapes perspectivism.

Except for Blake, the writers covered in this volume all accept, in one form or another, a kind of "learned ignorance," like that promulgated by Nicolas Cusanus. The human condition is frequently baffled by contradictions; we remain ignorant, but hope or believe that some higher power has the contradictions well in hand. Refusing to allow such ignorance, Blake blasts through the mysteries of contradictions to transcend, to transubstantiate, to transfigure the dull round of rationality. Stephen Behrendt describes Blake's break-throughs as "a daring end run around the reasoning intellect that is everywhere both the goal and the mechanism of Blake's art."[9] I prefer, however, to describe Blake's break-throughs as confrontations of the reasoning intellect with its contrary, the imagination. The reason, like Bacon, Newton, and Locke at the end of *Jerusalem*, is not to be defeated and abandoned, is not to be dismissed and avoided in an end run, but rather to be confronted at every turn, forced into illuminating coincidences of opposites. In fact, Behrendt's own description of Blake's method includes, confronts, and contains rationality instead of excluding it:

> Blake's art is grounded not in the empirical/analytical model of perception and cognition to which we have all become accustomed as readers and viewers but rather in one that invokes and exploits that conditioned expectation of empirical orientation in order to destablize and subvert it.[10]

Like deconstruction, with which Blake's method has many similarities, terms under erasure are questioned, fronted, opposed, but not completely destroyed. They still exist as vivid and necessary parts of the process of the imagination.

I can remember countless times when I dropped the books of the philosophers and picked up those of the poets again. The philosophers kept insisting that I abandon paradoxes, that I agree to submit to reasonable probability, step by logical step. But I always agreed with Blake, when he stated a similar preference for miracles over probabilistic history:

> Historians ... cannot see either miracle or prodigy; all is to them a dull round of probabilities and possibilites; but the history of all times and places, is nothing else but improbabilities and impossibilities; what we should say, was impossible if we did not see it always before our eyes.[11]

The coincidence of opposites is one of the major ways to celebrate those impossibilities and improbabilities that spring up every moment before us and in us.

The title of this volume, "All Nature is but Art" is taken from Pope's *Essay on Man,* in which the assertion means that man's limited viewpoint can not fully understand God's: according to Pope, what seems natural to us, perhaps even disorderly, is actually carefully planned by an intelligence much higher than ours. Nature, that which seems "natural," probably even inevitable, is actually a work of art created by God the artist. And if this is the best of all possible worlds, full in its divinely made plenitude, then the natural and the divine must be the same. When we see the universe from our limited viewpoint, we see a lack of coherent order. However, from God's point of view the universe is a unified work of art. Since man can never see from God's point of view, he can never fully understand the artistry in its wholeness. He must take the order of the universe on faith. But he can approximate or perhaps simulate an understanding of God's artistic wholeness by allowing his mind to entertain the existence of the coincidence of opposites. If he denies the coincidence of opposites, then he denies the possibility that God has created a work of art in the universe, for disqualifying the coincidence of opposites disqualifies the universe from being coherent. Nature, which can seem incoherent, disorderly, and even hostile to man, must be accepted as coherent, orderly, and congenial to man's higher purposes. Pope reconciles the traditional contraries of Nature and Art by insisting that man humble his reason to accept logical absurdities.

Although Blake's purposes in using the coincidence of opposites are quite different from Pope's, the motto "All Nature is but Art" can similarly epitomize his idea: what appears to us "given" and therefore natural is actually created, by man and his God working as one. At his most vehement, Blake condemns all unthinking ideology as grossly mistaken; in "London" in *Songs of Experience* the "mind-forged manacles," metaphorically created by the blacksmithing destructiveness of social consensus, bind us all without our knowledge or understanding. However, as always in Blake, the idea that "All Nature is but Art" contains its redeeming contrary as well: what appears to be natural is actually created by us, the Divine Humanity. Therefore, we need not simply accept and succumb to Nature, whether exhibited in the mores of society or in the stars of the heavens; we can, and must, create existence in each moment.

Many critics, implicitly or explicitly, dismiss any coincidence of opposites out of hand. The most extreme kind of dismissal comes from Morton Paley: in number two of his five principles of interpretation for Blake's illuminations in *Jerusalem* he declares,

> Interpretations according to which mutually exclusive meanings are seen as equally valid are not likely to be helpful. We should remember Blake's love of the definite.[12]

Paley's correct admonition—remember Blake's love of the definite—belies his advice. Love of the definite is the best way of achieving a vision of the coincidence of opposites. Only abstract reason denies that opposites co-exist. Living in definite minute particulars, which make up the impossible improbabilites of every moment's existence, could not happen if our rational minds stayed on guard to exclude acceptance of the coincidence of opposites.

Whether or not binary thinking is inescapably fundamental to human thought, Blake thought it was necessary to human existence, and if anyone has escaped its grasp (and I doubt that anyone has yet), the writers I study here, and of course I myself, are still inextricably caught in its grasp. Throughout this book I hope to show the productive contraries that can arise from binary thinking as Alexander Pope, Samuel Taylor Coleridge, Mary Shelley, Percy Bysshe Shelley, and William Blake struggle to conceive of and articulate a world that seems to tear itself apart in oppositions, oppositions that must coincide. Because without the extreme of total unity, we float isolated in the void; and without the extreme of complete individuality, we cluster indistinguishably in an amorphous lump. Both extremes are necessary to existence.

Notes

1. "1663 Butler Iliad 1.1.726 'For where the first does hap to be, The last does coincidere.'"

2. Indeed, in my natal chart, Venus and Mars are in conjunction, within 1.5 degrees of each other. I remember an astrologer telling me that such a conjunction of opposing forces creates a

"short circuit." Occasionally I still meditate on how to convert those opposing forces into dynamic reciprocity.

3. *The Marriage of Heaven and Hell,* E40. All quotations of Blake are taken from William Blake, *Complete Poetry and Prose of William Blake,* Ed. David V. Erdman (Doubleday, 1982). "E" with a number indicates the page in that edition.

4. J. Hillis Miller, "The Critic as Host," in *Deconstruction and Criticism,* ed. Harold Bloom (NY: Seabury Press, 1979), p. 252.

5. *The Romantic Ideology* (Chicago: Univ. of Chicago, 1983), pp. 1, 2.

6. *Romanticism and Gender* (NY: Routledge, 1993), p. 3.

7. Mellor, *Gender,* p. 4.

8. W.J.T. Mitchell, "Visible Language: Blake's Wond'rous Art of Writing," in *Romanticism and Contemporary Criticism,* eds. Morris Eaves and Michael Fischer (Ithaca: Cornell UP, 1986), p. 90.

9. Stephen Behrendt, *Reading William Blake* (NY: St. Martin's Press, 1992), p. 6.

10. Behrendt, *Reading,* p. 5.

11. "Descriptive Catalogue," E543.

12. *The Continuing City* (Oxford: Clarendon, 1983), p. 118.

"All Nature Is But Art"

Chapter 1
Nicholas of Cusa's *Coincidentia Oppositorum*

The English term "coincidence of opposites" is derived from the Latin term *coincidentia oppositorum*, the fullest exploration of which is attributed to Nicholas of Cusa, also called Nicolas Cusanus, whose 1440 *Of Learned Ignorance* (*De Docta Ignorantia*), sets forth a theological and geometrical justification for the concept. Like the Romantics in the eighteenth and nineteenth centuries, Cusanus in the fifteenth century felt discontent with the logical exclusions imposed upon him. Reacting against a theology that limited the power of God and the power of man, Cusanus set forth a counter theory. Similarly, reacting against a neo-classicism that limited the power of man, many Romantics accepted or assumed a theory similar to that of Cusanus. In response to the limits set by the rational compromises that guided the Enlightenment, Romantic writers sought a way to surpass exclusions by bringing together forces that neo-classical thought kept separate. Paradoxically, but appropriately, neo-classical writers also used the coincidence of opposites as a way of maintaining the *status quo*. But while Pope, the example of neo-classicism in this study, poses a mysterious coincidence of opposites that keeps man and God separate, Blake, and to a lesser extent the other Romantics, pose a coincidence of opposites that radically joins God and man.

Cusanus does not receive much attention from historians of philosophy; part of the reason can be found in the destructive nature of his theory, which denies a major aspect of the tradition of reason in philosophy. According to Pauline Moffitt Watts, Cusanus's doctrine of learned ignorance

> begins at that very point at which the usual modes of philosophizing collapse; the root of "learned ignorance" is the fact that absolute truth is beyond man's grasp. For this reason, Cusanus's new mode of speculation "undoubtedly vanquishes all modes of ratiocination of all philosophers."[1]

Ernst Cassirer puts the issue most succinctly: to reconcile opposites would dissolve philosophy itself.[2]

Because we in the twentieth century are so accustomed to an empiricist perspective, the very basis of the coincidence of opposites can easily elude us. The difficulty of acknowledging the concept even when the writer emphasizes it finds an unexpected example in Northrop Frye's posthumously published work *The Double Vision*.[3] Even though Frye, probably the best interpreter of Blake, takes his title from a poem by Blake, he interprets Blake's double vision in a way that gives a priority which Blake does not give and which the concept of the coincidence of opposites does not give. The phrase "double vision" occurs in a poem in a letter to Blake's patron Thomas Butts:

> For double the vision my eyes do see,
> And a double vision is always with me:
> With my inward eye 'tis an old man grey;
> With my outward a thistle across my way.
> (E721)

The poem goes on to see a living human in every object. In his explanation of this passage Frye comments, "the conscious subject is not really perceiving until it recognizes itself as part of what it perceives." This comment seems valid enough; it emphasizes the unity of subject and object, which is fundamental to the double vision which sees the coincidence of opposites. However, Frye goes on, "First, there is the world of the thistle, the world of nature presented directly to us."[4] A more careful reading of the quatrain reveals that Blake "first" describes the vision of his "inward eye," which perceives an old man grey, before he describes the vision of his "outward" eye. In a poem that emphasizes the constancy of his "double vision," Blake aids us in seeing with that double vision by first mentioning the half of his vision that is not ordinary sight, not what we would ordinarily, naturally expect to perceive first. Never, he claims, does he see with the single vision (later labeled in the poem "Newton's sleep,") that most of us would see and that Frye wants to put before us.

In Blake's double vision there is no such thing as "the world of nature presented directly to us"; such a vision is single vision, which is always already based on an ideology—in Blake's day, and in ours, usually an ideology of empiricism. Blake insists on always maintaining a double vision, in which a man and a thistle are seen at the same time and to the same extent. To help us to overcome the limits of single vision, Blake first describes the man seen by the inward eye. Despite that word order, however, and despite Blake's insistence on permanent double vision, Frye begins by interpreting the poem in terms of single vision. Such single vision seems most "natural" to us; Blake, and to a lesser extent the other writers considered in this book, fight to free us from that natural bondage. Blake's reversal of the ordinary way of seeing, an ordinary way of seeing that Frye himself falls into, implies a coincidence of opposites by its very defiance of common vision. Instead of interpreting this passage with the priority given by Blake, Frye interprets it with the priority given by ordinary vision, the very kind of vision that Blake is replacing (or reversing or extending) in his poem. Thus the simplest form of the coincidence of opposites, a basic reversal in priority between empirical vision and imaginative vision, becomes impossible to see with ordinary vision. Held prisoner by common sense, Frye over-rides even the most obvious implication of Blake's insistent message and surprising word order to impose the tyranny of single vision over the liberation of double vision. These comments are not meant to denigrate Frye but to illuminate how elusive the coincidence of opposites can be even to the most perceptive observer. Our habitual, limited ways of seeing make Blake either impenetrable or easily misconstruable.

With a similar singleness of vision, Leo Damrosch, one of Blake's finest interpreters, insists that proponents of the coincidence of opposites must accept things as they are. After explaining similarities between Blake and Cusanus, he implicitly consigns the coincidence of opposites to the conservative side:

> In the end Blake cannot truly reconcile contraries because, like the Puritans whose moral intensity he shares, he can only solve the problem of alienation by exclusion or casting out: sheep and goats. A true acceptance of opposites de-

mands a skeptical temperament and a commitment finally to things as they are.[5]

But how are things? Damrosch falls into a trap similar to the one that Northrop Frye falls into in *The Double Vision*: both Frye's "world of nature directly presented to us" and Damrosch's "things as they are" imply an empirical, common sensical seeing with the eye instead of an imaginative, Blakean seeing through the eye. "I question not my Corporeal or Vegetative Eye any more than I would Question a Window concering a Sight I look thro it & not with it."[6] Damrosch further loads his implicit dismissal of Blakean vision with the phrase, "true acceptance." What would be a "false" acceptance of the coincidence of opposites? According to Damrosch, Blake's "false" coincidence of opposites asserts unity by excluding recalcitrant elements. Instead, as I shall demonstrate in my chapters on Pope and Blake, it is Pope who excludes by accepting that there are things that we do not know, things that are known only by God. Blake, on the other hand, by insisting that man, like God, can know all, produces a "true" coincidence of opposites, one that does not accept the ideology of "things as they are."

Furthermore we probably should not look for Blake to "reconcile" contraries. When he defines the Prolific and the Devourer in *The Marriage of Heaven and Hell*, he insists:

> These two classes of men are always upon earth, & they should be enemies; whoever tries to reconcile them seeks to destroy existence.
>
> Religion is an endeavour to reconcile the two.
>
> Note. Jesus Christ did not wish to unite but to seperate [sic] them, as in the Parable of sheep and goats! & he says I came not to send Peace but a Sword.[7]

"'To believe in a God,' says Wittgenstein simply, 'means to see that the facts of the world are not the end of the matter.'"[8] Nor are they, in Blake's vision, even the beginning of the matter. I do not mean to belittle Frye or Damrosch; I adduce these examples to show how easy, how tempting, indeed how "natural" it is for us to fall into reading even Blake in conventional terms, even when he is explicitly, directly urging a larger vision upon us.

Both Pope and Blake insist on an identity between the universal and the particular, but they insist in very different ways.

Pope, more like Cusanus, uses the coincidence of opposites to justify mystery, to build a negative theology. Blake, extending Cusanus's implications, uses the coincidence of opposites to blast mystery, to build a constructive theology. My chapters on Pope and Blake will explore some of the details of this difference. Even though Pope uses the coincidence of opposites in a very conservative, mystifying way and Blake uses it in a very revolutionary, opening way, both of them include many strong elements of the opposite view. For example, Pope insists on the strength of the ruling passion, even though it apparently works against reason; Blake insists that the traditional God of Sinai inspired *Jerusalem*, even though most of the time he excoriates that God of mystery.

As so often in philosophical issues, the lines of argument in the coincidence of opposites can be traced back to Aristotle and Plato, or more precisely to the traditions promulgated by the followers of Aristotle and Plato. In general the position that opposites can coincide derives from Plato and his followers, especially Proclus the neo-Platonist, who especially admired Plato's dialogue *Parmenides*. (Cusanus himself commissioned a translation of this dialogue.) According to Ernst Cassirer, Cusanus more than any other thinker carried on the line of Plato: "Cusanus ... was perhaps the first Western thinker to attain an independent insight into the fundamental and essential sources of Platonic doctrine."[9]

The followers of Plato adhered to the law of unity: the Many is finally, ideally the One. This idealistic monism traditionally has been seen as less practical than Aristotle's more worldy viewpoint. Raymond Klibansky quotes the translator Georgius quoting Pope Nicholas V:

> Aristotle's political theory was more suitable to this life, while that of Plato was more appropriate to the state of innocence, had man not sinned and fallen.[10]

Coleridge makes a similar distinction between Plato and Aristotle:

> Plato's words are preparatory exercises for the mind. He leads you to see that propositions involving in themselves a contradiction in terms are nevertheless true; and which, therefore, must belong to a higher logic—that of ideas.

> They are self-contradictory only in the Aristotelian logic, which is the instrument of the understanding.[11]

A common sense viewpoint, one at home in the real world, refuses its opposite, an idealistic viewpoint, one at home in some other, better, imagined world. Coleridge, like Blake, sees through ordinary common sense to reverse that natural viewpoint.

The position that opposites can not coincide derives from Aristotle and his followers, who asserted the Law of Contradiction, also known as the Law of Non-Contradiction. Certain pairs of statements necessarily contradict each other and therefore can not both be true at the same time. This axiom is the most fundamental axiom of reasonable discourse, its *sine qua non*. Thus when Cusanus, and Pope, and Coleridge, and Mary Shelley, and Percy Bysshe Shelly, and Blake attack or even simply question the Law of Non-Contradiction, they are questioning not only Aristotle, but the very foundation of reason at its most necessary and self-evident base. As Aristotle asserts the axiom:

> the most certain principle of all is that about which one cannot be mistaken; for such a principle must be both the most familiar (for it is about the unfamiliar that errors are always made), and not based on hypothesis. For the principle which the student of any form of Being must grasp is no hypothesis; and that which a man must know if he knows anything, he must bring with him to his task. Clearly, then, it is a principle of this kind that is the most certain of all principles. Let us next state what this principle is. "It is impossible for the same attribute at once to belong and not to belong to the same thing and in the same relation"; and we must add any further qualifications that may be necessary to meet logical objections. This is the most certain of all principles, since it possesses the required definition; for it is impossible for anyone to suppose that the same thing is and is not, as some think Heraclitus says—for what a man says does not necessarily represent what he believes. And if it is impossible for contrary attributes to belong at the same time to the same subject (the usual qualifications must be added to this premiss also), and an opinion which contradicts another is contrary to it, then clearly it is impossible for the same man to suppose at the same time that the same

> thing is and is not; for the man who made this error would entertain two contrary opinions at the same time. Hence all men who are demonstrating anything refer back to this as an ultimate belief; for it is by nature the starting-point of all the other axioms as well.[12]

Among the many good explanations of Aristotle's idea is one by John Ferguson:

> Aristotle takes as the most certain of all principles the law of contradiction: 'the same attribute cannot at the same time and in the same respect both belong and not belong to the same subject' (1005b19). He feels so strongly on the importance of establishing this that he offers a series of proofs of its validity: for example, if all contradictory statements are true of the same subject at the same time, then all things will be one; the blunt fact is that all men do make some unqualified judgments, and Aristotle is content to stick to the common sense of that..... Aristotle denies that one opinion is as good as another, or that there is truth in appearances.... Those who deny the law of contradiction need to be convinced that there is an unchanging reality, something which is prior to sensation.... the alternative is a thoroughgoing relativism, which Aristotle regards as absurd. (1011b23).[13]

We begin, then, in Aristotle's Law of Non-Contradiction, from a bedrock of common sense, without which the entire universe of human epistemology would supposedly collapse. As Aristotle's ideas were extrapolated and discussed through the centuries, and as the science of logic developed, hundreds of elaborate and intricate schemes, replete with diagrams and the special language of the cognoscenti, found their place in the ongoing discussion. Here is a typical example of a twentieth-century textbook exegesis:

> Standard-form categorical propositions having the same subject and predicate terms may differ from each other in quality [affirmative or negative] or in quantity [all, some, no] or in both. This kind of differing was given the technical name 'opposition' by older logicians, and certain important truth relations were correlated with the various kinds of opposition. Two propositions are *contradictories* if one is

> the denial or negation of the other, that is, if they cannot both be true *and* they cannot both be false. It is clear that two standard-form categorical propositions having the same subject and predicate terms but differing from each other *both* in quantity and in quality are contradictories.... The traditional or Aristotelian account of categorical propositions held that universal propositions [**A** and **E**] having the same subject and predicate terms but differing in quality were contraries.... Schematically we may say that the contradictory of "All S is P" is "Some S is not P," and the contradictory of "No S is P" is "Some S is P"; **A** and **O** are contradictories, as are **E** and **I**.[14]

Whether contraries or contradictories (we shall meet with some similar, but more crucial distinctions later in Coleridge and Blake), the principle of Non-Contradiction stands: logically, when contraries or contradictories exist, one side completely, irrevocably, logically, common sensically excludes the other. They cannot both be true.

So powerful did Aristotle's Law of Non-Contradiction become in medieval Christianity that it limited even the power of God: He could not do anything that would violate that law. The general philosophical problem of whether God could contain logical contradictions sometimes found expression in one particular question: can God foresee the future? If He can, then man does not have free will. It is not possible for God to foresee something and for it not to happen. According to the Law of Non-Contradiction, we cannot have it both ways at the same time. Likewise, it would be impossible for God to foresee that something will not happen and for that thing to happen. Thus God's foreknowledge and man's free will come into philosophical conflict, as a subsection of the problem of non-contradiction even within the omnipotence of God.

Cusanus's coincidence of opposites provides a way out of that dilemma. One of his answers expands the concept of possibility:

> human nature embraces both those who are and those who neither are nor will be, though they could have been. Consequently, if even what never shall be should come to pass, nothing would be added to divine Providence, for it equally comprises actual events and those which, though possible, do not take place.[15]

Thus Cusanus's coincidence of opposites includes not only what happens or what might happen; it includes both what will happen and what will not happen. We can have it both ways. In fact, we must have it both ways. In God's power, the old Aristotelian Law of Non-Contradiction becomes replaced by the coincidence of opposites. The law of limited possibilities is transcended, as is the law that two things cannot happen at the same time if they are logically inconsistent. For in the timeless mind of God, which contains all past, present, and future, all possibilities and all non-possibilities must exist, including the improbable and the supposedly impossible.

Cusanus further develops his refutation of God's limitations that would be imposed by the Law of Non-Contradiction:

> God's infinite Providence ... embraces what shall happen and what, though possible, shall not happen; and, in much the same way as a genus contains different species, it includes things contrary to one another.... What God has foreseen He necessarily has foreseen, for His Providence is necessary and unchangeable; yet He was also able to foresee the opposite.... Though tomorrow, e.g., I can choose between reading and not reading, my choice, whichever it be, is known to Providence, for contraries are among its objects. Therefore, whatever I shall have chosen to do will be done in accordance with divine Providence.[16]

This kind of double talk clearly does not please the rational philosopher. J.B. Hawkins, in his introduction to Germain Heron's translation of *De Docta Ignorantia*, gently mocks:

> There are not many who are likely to take all this very seriously as an adumbration of ultimate mystery. Some will be irresistibly reminded of Lewis Carroll, and others may despise what seems to be no more than conjuring with words.[17]

Certainly Cusanus is conjuring with words, but as a clever conjuror he produces a magic that joyfully exploits the potential of language to reveal and even embody the philosophical abstractions that fall short of explaining life.

Cusanus's coincidence of opposites frees man from the tyranny of the senses, the tyranny of common sense, the tyranny of logic.

By freeing God from lower logic, Cusanus also frees man into a closer apprehension of his divinity:

> the *coincidentia oppositorum* represents more than Cusanus's critique of scholastic logic and natural philosophy. Through it, Cusanus also transcends the late scholastic notion of the *potentia absoluta* of God; namely that God, through his absolute power, could do or make anything provided it did not violate the law of contradiction. For Cusanus, the law of contradiction itself qualifies God's freedom and omnipotence.[18]

Thus Cusanus does, as will Blake, use the coincidence of opposites as a way of freeing God from human constraints. Blake, by combining human and divine, will also free man from such a limited view of the divine humanity. Both Blake and Cusanus see God as more free than tradition allows:

> in conceiving of God as the *coincidentia oppositorum*, Cusanus went beyond all of the scholastics, and established more radical guarantees for the absolute freedom and transcendence of the divine.[19]

Blake then goes further to establish yet more radical guarantees for the absolute freedom and transcendence and immanence of the divine humanity.

Ernst Cassirer, in very Coleridgean language, puts Cusanus firmly into the history of philosophy when he insists on the central role that Cusanus played in the Renaissance:

> Any study that seeks to view the philosophy of the Renaissance as a *systematic* unity must take as its point of departure the doctrines of Nicholas Cusanus. Of all the philosophical movements and efforts of the Quattrocento, only his doctrines fulfil Hegel's demand; only they represent a 'simple focal point' in which the most diverse rays are gathered. Cusanus is the only thinker of the period to look at all of the fundamental problems of his time from the point of view of *one* principle through which he masters them all.... [That principle] includes the totality of the spiritual and physical cosmos.[20]

As we shall see later, Coleridge insisted on a similar unified system.

Cusanus's purpose in his *coincidentia oppositorum* could be said to be to justify the ways of God to man, that is, to re-write man's perspective into God's perspective. From man's perspective, from the perspective of Aristotelian logic, man and therefore God are both limited. But from God's perspective, according to Cusanus and the coincidence of opposites, God is not limited. The important thing for Cusanus is not the coincidence of opposites per se, but rather the infinity of God. The coincidence of opposites is a tool for realizing, or at least accepting, that infinity. Similarly, Blake will use the coincidence of opposites as a tool for realizing the infinity and unity of God and man. Although Cusanus keeps an infinite distance between man and God, Blake will bridge that distance.

The path taken by Cusanus to reach his new concept of the freedom of God in the *coincidentia oppositorum* begins with two opposing, complementary principles: the principle of expansion and the principle of limitation. The principle of expansion states that man, like all created things, wants to find his fullest possible self-expression:

> We see that God has implanted in all things a natural desire to exist with the fullest measure of existence that is compatible with their particular nature.[21]

In his exploration of the concept of the Great Chain of Being, A.O. Lovejoy calls this principle the principle of plenitude.[22] At the same time, in the opposite direction, this striving toward fullness is limited by the particular position of the creature. In man's case, his inquiry must begin with and is limited to what he knows. These two opposing forces act as centripetal and centrifugal forces, at once thrusting man out as far as he can go to his potential and at the same time limiting him within his orbit.

Cusanus thinks that man must begin with what he knows from his limited perspective. Beyond what he knows, he can only analogize: "every inquiry is comparative and uses the method of analogy."[23] As we shall see, this latter principle of limitation is the keystone to Alexander Pope's whole theory of man's knowledge of the universe. Here in Cusanus, the two opposing tendencies work equally. Cusanus's principle of limitation is based on a concept of infinity as beyond human understanding and therefore beyond all analogizing: "the infinite as infinite is unknown, since it is away and above all comparison."[24] Even the greatest possible expansion, the fulfillment of man's capacities, no matter how close it might theoretically come, can not, by its finite nature, achieve or comprehend the infinite. Man's limitation must finally over-rule his potential for expansion.

As we shall see later, Blake denies this inaccessibility of the infinite to man. In particular in the tractates he insists that man could not desire infinity if he did not perceive it:

> More! More! is the cry of a mistaken soul, less than All cannot satisfy Man.... The desire of Man being Infinite the possession is Infinite & himself Infinite.[25]

Therefore perception must exist beyond the senses, and by implication beyond the ordinary limitations as posed by Cusanus and Pope.

Part of man's limitation resides in language, which embodies the problem of specificity. Just as we must, according to Cusanus, rise above our physical human restrictions, so we must transcend language, which in its abstractions allows us to formulate the Law of Non-Contradiction. But in Blake's poetry, language is used not only as a trap in the Law of Non-Contradiction, but also as a means of escape. Cusanus sees language as an absolute limit to the potential of man's understanding of God. Language must therefore be transcended, risen above. Blake's insight is that this kind of transcendence can be blended with an absolute immanence. Thus language, while remaining limited by the real world, at the same time is the means of transfiguration into the visionary world. Cusanus wants to deny the potential of language. He values "rising above the literal sense of the words ... leaving aside what is sensible in them in order to arrive unimpeded at what is purely intelligible."[26] Cusanus's method thus sounds like traditional Platonism: "the more we abstract from sensible conditions, the more certain and solid our knowledge is."[27] Cusanus's method leaves behind the empirical as a standard. He is not searching for something material. He is rather searching for something beyond the material. To man's understanding such a goal is nothing because it is not accessible to the senses: "an understanding of God is not so much an approach towards something as towards nothing."[28]

This negative theology, this insistence by Cusanus on wordlessness, of believing in a realm beyond articulation even if we can never experience it directly, is one of the main streams of his neo-Platonism. The tradition of Parmenides always leads to this wordlessness:

> In analogy with this movement of the Soul towards the One, Parmenides removes at the end of the discussion, not only all affirmation, but also all negation and, indeed, all speech, indicating that all discourse about the One leads to a point where language fails.... The last step is to purge the Soul from all dialectic activity. Thus, by negation, Parmenides removes negation itself. And silence ends the discourse of the One.[29]

Of course such silence would remove all language, all writing, all literature. That which Cusanus and Pope claim to be beyond the realm of language, the coincidence of opposites, Shelley and Blake will manifest in language.

Cusanus's denial of language is based on his search for a deeper truth, an essence. He believes that there is a bedrock truth, a Platonic Christian God, whose fundamental nature is by definition unavailable to human understanding: "the quiddity of things, which is ontological truth, is unattainable in its entirety."[30] Even though the method of attempting ignorant understanding of the existence, essence, and ways of God must proceed from the known in hopes of acknowledging the unknown, ultimately such a goal can never be reached: "infiniti ad finitum proportionem non esse."[31]

Man's limited understanding can never grasp the totality because of its very nature: "what man observes in nature is absolute incommensurability.... Man's reason operates only within the realm of inequalities and contradictions."[32] According to Cusanus, there is no proportion between the infinite and finite because any proportion needs similar qualities.[33] Man's limited abilities mean that he can not by reason or understanding alone reach the ultimate truth: "understanding ... is fundamentally unable by any rational process to reconcile contradictories."[34]

In fact Cusanus even uses man's reason to link man to the natural level, the level that he shares with the rest of the material, lower creation. Instead of reason being the highest quality of man, it is rather analogous to the instinct of animals: "Reason, Cusanus says ... where he quotes Jerome quoting Philo, is a capacity that man shares with animals. Man uses logic in much the same way that a hunting dog instinctively uses tracks in searching for his prey."[35] This idea is very similar to Blake's: reason is not the highest capacity of man, and certainly is not capable of leading him to God. It is rather part of his natural self, just like his body.

From the earthly perspective the totality of God can never be achieved. Like the infinite regression of so many logical systems we can approach, but never attain:

> our intellect, which is not the truth, never grasps the truth with such precision that it could not be comprehended with infinitely greater precision. The relationship of our intellect to the truth is like that of a polygon to a circle; the resemblance to the circle grows with the multiplication of the angles of the polygon; but apart from its being reduced to identity with the circle, no multiplication, even if it were infinite, of its angles will make the polygon equal the circle.[36]

Despite this use of geometrical analogy to disqualify man's understanding from reaching God's, Cusanus will develop this analogy of the relationship of polygon to circle into his primary visual image for the coincidence of opposites as he imagines the infinite polygon becoming the infinite circle, or the infinite triangle becoming the infinite line. The geometry which is here used to analogize the impossibility of our ability to reach infinity is used in other places to take us as close to that realm as possible. The reach is still logically, humanly impossible, and the mathematics still shows the infinite distance between human and divine, which can not be bridged by merely logical, human means, but only by the coincidence of opposites.

Cusanus bases his combination of theology and geometry on certain axioms, perhaps the most fundamental of which is his assumption about infinity:

> It is impossible to have more than one infinite.... the infinite triangle, though it is the perfect model of all triangles, is not composed of a plurality of lines, is not in any sense a compound, but is more perfectly indivisible; and since it is

> the perfect model triangle, it must have three lines; therefore, the one infinite line is itself three lines and these three lines are one perfectly indivisible line.[37]

These assumptions and conclusions about the infinite lead inevitably to a coincidence of largest and smallest:

> since each part of the infinite is infinite, then one foot of the infinite line is as equally the whole infinite line as two feet are.[38]

We are coming very close to seeing Blake's world in a grain of sand.

Cusanus's geometry begins with the principle of non-contradiction in the finite world; different geometrical figures must remain different:

> elementary figures [circle, triangle, sphere] etc., which are the only ones educible from the potency of the simple line, cannot be reduced to one another in the finite.[39]

In the infinite realm, however, these figures can coincide:

> The infinite circle has infinite diameter and circumference.... the centre, diameter and circumference are one and the same.... the centre is the circumference.[40]

Thus, through the logic of geometry raised to the power of infinity, Cusanus arrives at the traditional mystical formula that the center and circumference come together. For, in the finite realm, the polygon can never become a circle, no matter how many angles it contains; in the finite realm, a triangle can never can become a straight line, no matter how long its sides become; in the finite realm, a circle can never become a straight line, no matter how long its diameter becomes. But in the infinite realm, the infinite polygon does coincide with the infinite circle, the infinite triangle does coincide with the infinite straight line, and the infinite circle does coincide with the infinite straight line. By picturing the largest possible polygon stretched out beyond the possible into infinity, we can, by analogy to our finite world, begin to comprehend the infinite world. With a bit more imaginative difficulty, we can similarly begin to comprehend the expansion of the infinite circle into a straight line. We can stretch it out beyond our comprehension to acknowledge infinity by analogy to what we can comprehend.

Our human comprehension, although capable of imagining analogies to the infinite, can never grasp the infinite. Thus, by analogy to what we know, we can describe or accept or tolerate what we don't know, what we can never know. Pope will take an approach similar to Cusanus's: we must remain content with partial knowledge, even while we understand as much as we can through the reasoning power of analogy. We can attain only learned ignorance. Blake, however, will insist that a coming together of finite and infinite, similar to that of Cusanus, guarantees total knowledge to the human, who participates fully in the divine. Both Cusanus and Pope want to start from what we know and then proceed logically up to the limits of our knowledge, accepting but never attaining totality. Blake, however, will make the leap, based not on logic but on miracles.

According to Cusanus and Pope, in the other realm, beyond our knowledge, the ordinary laws of non-contradiction do not apply. Even the distinction between the largest and the smallest must fail to operate in the infinite mind of God:

> the absolute maximum is one and it is all; all things are in it because it is the maximum. Moreover, it is in all things for this reason that the minimum at once coincides with it, since there is nothing that can be placed in opposition to it.[41]

Insofar as the finite limits of man cannot reach the infinite and insofar as the Maximum must coincide with the Minimum because of the very nature of infinity, a possibility of joining the two incommensurate realms occurs. The maximum possibility of the infinite is the very joining of infinite and finite. As in Blake, the only possible earthly existence of such a principle of maximum/minimum is in the divine made human, or spirit made flesh:

> the one maximum in which the universe finds especially and most completely its actual and ultimate subsistence.... this maximum bears the ever blessed name of Jesus.[42]

Dorothea Singer, while writing primarily about Giordano Bruno, links infinity, the coincidence of opposites, and Jesus all together because each of the three implies the other two:

> In both writers [Bruno and Cusanus], closely associated with belief in the infinity of the universe was the doctrine of the Coincidence of Contraries. The subject-object relationship similarly was envisaged by both writers as a process of admixture culminating in identity.... Cusanus gave the doctrine a new slant and a new emphasis.... Cusanus saw Salvation as the Line of Unification between Contraries.[43]

Salvation through Jesus, for Cusanus, and for Blake, is the way to unify the contraries of infinite and finite:

> The universe is completely dependent upon God but remains in an essential antithesis to him as the limited to the infinite, the relative to the absolute. How is this antithesis to be overcome? ... The perfect mode of unity would be a being which was both creator and creature and in which the relative, without ceasing to be relative, would be one with the absolute, *maximum contractum* and *maximum absolutum* at the same time (III,2).... Nicholas embraces the thesis of the Scotist school of theologians that the perfection of creation calls in some sense for the Incarnation.[44]

Like Blake, Cusanus finds the solution to the logical problem in the person of Christ:

> As in Cusanus' theology of Jesus as the universal Form, Blake asserts an immanent God who is nonetheless divine. 'All deities reside in the human breast' is a formulation intended to emphasize the divine element in man rather than to demote God to a metaphor of individual human imaginations; it is in God that the individual is freed from isolation and solipsism.[45]

Cusanus's inquiry logically leads to negative theology, in which affirmations serve no purpose. Since affirmations can state only human knowledge, divine knowledge must be stated in negative terms. "Since a term that is particular, that marks a distinction and that suggests its opposite, can only apply to God in the way we have described, affirmations, as Denis says, are unsuitable."[46]

Cusanus's theory thus fits into the Jewish tradition of allowing no graven images of God because any image is an affirmation, which necessarily distorts the truth:

> The pagans of old ridiculed the Jews for adoring a God unique and infinite whom they did not know, whilst they themselves were adoring Him in his manifestations, adoring Him, in other words, wherever they beheld His divine works.[47]

Such glorification of the physical can fall into a kind of materialism, which would never allow a coincidence of opposites. Consistent throughout the tradition of the coincidence of opposites, from Plato through Cusanus to Blake, is an insistent stand against letting materialism determine the limits of human or divine. "Negative Theology ... is so indispensable to affirmative theology that without it God would be adored, not as the Infinite but rather as a creature, which is idolatry."[48] This necessity of believing in abstraction will be followed by Pope but vehemently rejected by Blake. Pope and Cusanus take a turn that Blake does not take into Negative Theology. For Cusanus, a denial of materialism, a denial of the Law of Non-Contradiction can not be an affirmation. As we shall see, for Blake it can.

"All affirmations, therefore, that are made of God in theology are anthropomorphic, including even those most holy Names ... used by the Hebrews and the Chaldeans."[49] To Cusanus, anthropomorphizing of God limits Him. Just as Cusanus would not allow God to be limited by any restrictions on his knowledge of the future and the allowing of free will, so he will not allow the limits of God reduced to man. But anthropomorphism to Blake is not something to avoid; on the contrary it is the very nature of the Human Form Divine; the divine implies, even requires the human, and vice versa.

Apparently none of the elements of Cusanus's philosophy is original. Cusanus's originality is in putting them all together. Such a judgment is very appropriate for the philosophy of the coincidence of opposites, since it shows how disparate elements can be fitted together to form a new whole:

> The general plan of the book is not very original; it follows the customary Neoplatonic scheme of the outflow of things from God and their return to him which had been represented in the earlier middle ages, for example, by the *De Divisione Naturae* of John Scotus Erigena. The notion of the reconciliation of contraries in God, the coincidence of oppo-

> sites, is fairly evidently derived from Eckhart. The theory of enlightened ignorance, the *docta ignorantia* itself, is only a new expression and extension of the *theologia negativa* which was familiar enough to all mediaeval thinkers, especially as found in the writings of the Pseudo-Dionysius.
>
> Nicholas does not tell us what precisely he regarded as his special inspiration, but we might well suppose that it was the way in which these elements fitted together, for it is in this respect that he is really original.[50]

Perhaps not coincidentally the idea in its completeness seems to have come to him while he was engaged in his ambassadorial duties of reconciling opposing factions. Although Cusanus's life as such is not essential to understanding the coincidence of opposites, he did develop the same theme in theory and in practice. In his work as papal ambassador, he often tried to solve problems of papal schism.[51]

Jasper Hopkins points out many other lines of thought from which Cusanus may have been working, as he claims that Cusanus's work in *De Docta Ignorantia* was a synthesis in the debate between Ockhamism and Thomism, nominalism and realism:

> In it can be found the influence of Eckhart, of the Hermetic tradition, of Pseudo-Dionysius, and of Boethian mathematics, together with a newly devised cosmological framework and a newly conceived theology of redemption.[52]

Cusanus himself most often acknowledges the influence of Dionysius the Areopagite, whose "disjunctive" and "copulative" theology he praises and imitates:

> Cusanus indicates in *De Docta Ignorantia* that he has derived his conception of God as the *coincidentia oppositorum* from the negative theology of thinkers such as Dionysius the Areopagite. He refers specifically to the *Mystical Theology*, *On Divine Names* and letter to Gaius. Cusanus has then substituted a Platonic-Dionysian metaphysical dialectic based on the opposition between the finite and the infinite for the scholastics' dialectics of logic and discursive reasoning.[53]

Jasper Hopkins, who has translated and commented on Cusanus's major works, often seems to want to tame Cusanus, to reduce the coincidence of opposites by offering a logical critique of it:

> a number of Nicholas's statements appear, prima facie, to be unintelligible. The most noteworthy—and, ironically, the best known—is his assertion that God is a sphere whose center is everywhere and whose circumference is nowhere. This statement ... is not original with Nicholas but derives from Pseudo-Hermes Trismegistus's 'Book of the Twenty-four Philosophers,' a compilation of the late twelfth or the early thirteenth century. Nicholas, however, seems to have borrowed this formula from Meister Eckhart, who uses it in no less than six places. Yet, Eckhart's statement—already bizarre in its attempt to express the doctrine of divine omnipresence—becomes all the more bizarre in the mouth of Nicolas, who comes to declare that even the *machina mundi* has its center everywhere and its circumference nowhere. Not only is it unclear what sense it would make to apply this rubric to the mechanism of the world, but the reason offered in support of so applying the rubric appears unintelligible: vis., that God is the world's center and circumference—God, who is present everywhere and nowhere. But the sense in which God, an immaterial being, could be the world's center and circumference is never successfully elucidated.[54]

This kind of literal-minded interpretation makes Cusanus sound even beyond nonsense. When Cusanus adopts the beautiful traditional geometrical formula of the mystics that God is a sphere whose center is everywhere, he is of course emphasizing a point that Blake will emphasize in different language: the relationship between God and human beings is such that God exists in each and every point equally. Traditional theologies see God either as distant or as all-encompassing in the nature of a circumference. Thus God is large and unfathomable. In the coincidence of opposites God exists in every minute particular; His center is not only out there somewhere but is also always in here, a maximum in the minimum. Blake will similarly bring together divine and human and develop Cusanus's implications.

More satisfactory than Hopkins's attempts to argue away the contradictions in Cusanus is the concept of a copulative theology (as mentioned above in Cusanus's homage to Dionysius). Copulation suggests the joining of two different dimensions, which come together, not through compromise, but through intercourse, interpenetration.

Another way to define copulative theology comes from plant imagery, like that of Coleridge's organicism:

> The human mind … is a divine seed that comprehends in its simple essence the totality of everything knowable; but in order for this seed to blossom and to bear fruit, it must be planted in the soil of the sensible world. The basic character of that 'copulative theology' sought by Cusanus lies in this reconciliation of mind and nature, of intellect and sense.[55]

> For Dionysius the Areopagite, 'deification' … takes place according to the hierarchical principle, i.e., in a completely determined series of steps.... for Cusanus it is a single act, one in which man puts himself into an immediate relationship to God.[56]

Here Cusanus's divergence from the tradition that he is following again anticipates Blake, who also dismisses a step-by-step, logical process as the way to enlightenment. In his tractates, Blake parodies the style of logic to use logic to refute this very idea:

> II Man by his reasoning power. can only compare & judge of what he has already perciev'd.
>
> III From a perception of only 3 senses or 3 elements none could deduce a fourth or fifth.
>
> IV None could have other than natural or organic thoughts if he had none but organic perceptions
>
> I Mans perceptions are not bounded by organs of perception. he percieves more than sense (tho' ever so acute) can discover.
>
> …
>
> Application. He who sees the Infinite in all things sees God. He who sees the Ratio only sees himself only.
>
> …

> Therefore God becomes as we are, that we may be as he is.[57]

Blake thus develops a notion of God like that of Cusanus, but brings it to a higher level. Blake's man is not limited by the five senses and the rational mind. Blake's man is capable of leaping to insight directly because he perceives more than limited senses can perceive.

The best twentieth-century explanation I have yet seen of the kind of process that Cusanus, and even more so Blake, are trying to accomplish occurs in *Philosophy of the Literary Symbolic*, by Hazard Adams. Basing most of his literary theory on Blake (although Adams insists on maintaining a secular approach instead of Blake's religious one) he defines a coincidence of opposites that captures precisely the kind of inclusiveness that I am trying to elucidate. In offering to replace the viewpoint of deconstruction that all language is duplicitous, Adams asserts:

> I choose to see another side—secular creative expression—and claim that from poetry's point of view the poem makes a language, which creates and contains its signified and allows it to emanate into the world to be devoured; though I shall recognize that from the opposing point of view, which I shall call 'antimyth,' it copies or signifies only and cannot contain or radically form. And I shall hold that both views are necessary fictions—Blakean contraries.... both myth and antimyth are, as I apply the terms to Blake, modes of imaginative construction that result in what may be called fictions.... Because both are modes of activity and not themselves copies of anything, the question of which has truth or correspondence to reality is not a possible question; indeed, the question is merely a reflection of a category, to borrow Kant's term, of the antimythic mode, and is meaningful only within its terms.[58]

Thus Adams gives us redefinitions of the coincidence of opposites and, incidentally, answers Hopkins's misunderstandings or reductions of Cusanus. Instead of trying to assert absolute Truth, Blake (or at least Adams's theory derived from Blake) lives in creativity and dialogue instead of in tyrannical Truth. The relation between fallen vision and Visionary Vision is a dynamic one, even though only the latter, not the former, can be aware of that dynamism:

> without myth, antimyth starves. In this sense, myth potentially contains antimyth (as the seed does the plant): antimyth can never contain myth, though it is engaged in a constant effort to devour it.[59]

Most of all, Adams gives us a way to view language that allows its limitations, but at the same time recognizes its redemptive power:

> The stubbornness of the structure is two-fold: From the 'allegoric' point of view, language has the capacity to resist the stasis reason desires and stubbornly frustrates those who—like Urizen, Satan, and the 'priesthood'—would choose to abstract a single form of worship from it (try to reduce it to a system pointing outward only to that one form—the clock world of deism or the ideal world of Platonism). This capacity to resist is its mythical, Los-like structure. On the other hand, language resists stasis only as a result of the struggle that Los has with its equally dangerous (from the poet's point of view) but necessary antimythical susceptibility to externalization and hardening. Its stubbornness, then, can point either way.[60]

Cusanus, in his anticipations of Romantic theory, comes very close to Blake's and Coleridge's definitions of the creative imagination. Pauline Watts explains that, while discussing God's creation of man in his own image Cusanus finds

> the transcending principle for which he has been looking and establishes the grounds for a positive theology and anthropology.... Man, Cusanus asserts, creates the conjectural world in a likeness of the manner in which God has created the 'real' world. Metaphor is no longer however the obfuscated and obfuscating product of man's disjunction from God and the universe. Metaphor now is the tool that man uses to create his conjectural world, the proof of his Godlike nature. This formulation is an affirmative anthropomorphic principle which counteracts the impersonal disjunctive metaphysic that dominated ... *De Docta Ignorantia* because it stresses the essentially creative and active nature of both

> God and man. God has created the universe not out of necessity but freely. Correspondingly, man, the image of God, is the creator of his conjectural and metaphorical universe. Man is no longer a stranger in a universe that he can never know directly. The fact that he is the image of God and the creator and ruler of his own universe is the source of his cosmic status and dignity.[61]

Thus the coincidence of opposites leads to a creative definition of human existence.

Much of what Cusanus helps us understand about Blake is also helped by Carl Jung's formulations of religious consciousness. In his *Answer to Job* Jung, like Cassirer, uncovers the centrality of the coincidence of opposites to medieval thought:

> the study of medieval natural philosophy—of the greatest importance to psychology—made me try to find an answer to the question: what image of God did these old philosophers have? Or rather: how should the symbols which supplement their image of God be understood? All this pointed to a *complexio oppositorum* and thus recalled again the story of Job to my mind: Job who expected help from God against God. This most peculiar fact presupposes a similar conception of the opposites in God.[62]

This concept of God against God helps to answer the earlier bafflement of Hopkins about how humanity can be both divine and not divine. Just as God is by necessity a coincidence of opposites, so is man in his divine humanity, for and against the self.

> This is perhaps the greatest thing about Job, that, faced with this difficulty [a God who does not care for moral opinion] he does not doubt the unity of God. He clearly sees that God is at odds with himself—so totally at odds that he, Job, is quite certain of finding in God a helper and an 'advocate' against God. As certain as he is of the evil in Yahweh, he is equally certain of the good. In a human being who renders us evil we cannot expect at the same time to find a helper. But Yahweh is not a human being: he is both a persecutor and a helper in one, and the one aspect is as real as the other. Yahweh is not split but is an *antinomy*—a totality of inner opposites—and this is the indispensable

> condition for his tremendous dynamism, his omniscience and omnipotence.[63]

This double nature of God is essential for understanding Cusanus and later Shelley and Blake. There is a doubleness at the very heart of existence, in the person of God and the very ontology of humankind. For that reason, as for example in *Jerusalem* plate 43, one cannot separate out the bad and assign it unmistakably against the good. The good and the bad must always be united, in an apparently impossible conjunction, a disjunctive copulation of incommensurables.

> The paradoxical nature of God ... tears [man] asunder into opposites and delivers him over to a seemingly insoluble conflict.... The doctor therefore advises his patient to wait and see.... As experience shows, symbols of a reconciling and unitive nature do in fact turn up in dreams, the most frequent being the motif of the child-hero and the squaring of the circle, signifying the union of opposites. Those who have no access to these specifically medical experiences can derive practical instruction from fairy tales, and particularly from alchemy. The real subject of Hermetic philosophy is the *coniunctio oppositorum*.[64]

In Jung's psychology, dreams are the main form of communication from the unconscious realm, where opposites can coincide, to the conscious realm:

> all unconscious nature longs for the light of consciousness while frantically struggling against it at the same time. The conscious realization of what is hidden and kept secret certainly confronts us with an insoluble conflict; at least this is how it appears to the conscious mind. But the symbols that rise up out of the unconscious in dreams show it rather as a confrontation of opposites, and the images of the goal represent their successful reconciliation.[65]

Similarly, in the poetry of the writers studied here, images arise from the realm of the *coincidentia oppositorum* into the realm of language, where poetry links the two realms together. For Cusanus and for Blake, the ultimate such link requires a God-man:

> Eschatology means in effect that Christ is God and man at the same time and that he therefore suffers a divine as well as a human fate. The two natures interpenetrate so thoroughly that any attempt to separate them mutilates both.[66]

Blake's human form divine.

Notes

1. Pauline Moffitt Watts, *Nicolaus Cusanus: A Fifteenth-Century Vision of Man* (Leiden: E.J. Brill, 1982), p. 43.

2. Ernst Cassirer, *The Individual and the Cosmos in Renaissance Philosophy*, Trans. Mario Domandi (NY: Harper, 1963), p. 17.

3. *The Double Vision: Language and Meaning in Religion*. Univ. of Toronto Press, 1991.

4. Frye, *Double*, p. 23.

5. Leopold Damrosch, *Symbol and Truth in Blake's Myth* (Princeton UP, 1980), p. 242.

6. "A Vision of the Last Judgment," E566.

7. *The Marriage of Heaven and Hell*, plates 16–17, E40.

8. Notebooks 1914–16, p. 74e; qtd. in Thomas McFarland, *Coleridge and the Pantheist Tradition* (Oxford: Clarendon, 1969), p. 55.

9. Cassirer, p. 15.

10. Raymond Klibansky, "Plato's *Parmenides* in the Middle Ages and the Renaissance: A Chapter in the History of Platonic Studies," *Medieval and Renaissance Studies* 1 (1943), p. 300.

11. *Table Talk*, 30 Apr 1830; qtd. in Kathleen M. Wheeler, *Sources, Processes and Methods in Coleridge's* Biographia Literaria (Cambridge: Cambridge UP, 1980), p. 49.

12. *Metaphysics*, trans. Hugh Tredennick (Cambridge: Harvard UP, 1936), pp. 162–63; 1005b13–34.

13. John Ferguson, *Aristotle* (NY: Twayne, 1972), pp. 113–14.

14. Irving M. Copi, *Introduction to Logic*, 7th ed. (NY: Macmillan, 1986), pp. 177–8.

15. Nicolas Cusanus, *Of Learned Ignorance*, trans. Germain Heron (New Haven: Yale UP, 1954), p. 49.

16. Cusanus, p. 50.

17. J.B. Hawkins, "Introduction," in Cusanus, p. xix.

18. Watts, p. 46.

19. Watts, p. 227.

20. Cassirer, p. 7.

21. Cusanus, p. 7.

22. Arthur Lovejoy, *The Great Chain of Being*, 12th ed., (Cambridge: Harvard UP, 1936), p. 52 and *passim*.

23. Cusanus, p. 7.

24. Cusanus, p. 8.

25. E3.

26. Cusanus, p. 10.

27. Cusanus, p. 25.

28. Cusanus, p. 39.

29. Klibansky, p. 288.

30. Cusanus, p. 12.

31. "There exists no proportion between the finite and the infinite"; *De Docta Ignorantia* [*Of Learned Ignorance*], qtd. in Carlo Riccati, *"Processio" et "Explicatio": La doctrine de la création chez Jean Scot et Nicolas de Cues* (Naples: Biblipolis, 1983), p. 27.

32. Watts, p. 44.

33. Dana Bliss Birkby, *Salting the Tail of the Infinite: Mathematics in the Epistemology of Nicholas of Cusa*, unpublished thesis (Harvard, 1982), p. 14.

34. Cusanus, p. 13.

35. Watts, p. 44.

36. Cusanus, p. 11.

37. Cusanus, p. 31.

38. Cusanus, p. 36

39. Cusanus, p. 45.

40. Cusanus, p. 47.

41. Cusanus, p. 9.

42. Cusanus, p. 10.

43. Dorothea Waley Singer, *Giordano Bruno: His Life and Thought* (NY: Henry Schuman, 1950), p. 80.

44. Hawkins, p. xxv.

45. Damrosch, p. 247.

46. Cusanus, p. 55.

47. Cusanus, p. 58.

48. Cusanus, p. 59.

49. Cusanus, p. 56.

50. Hawkins, p. xiv.

51. Birkby, p. 5.

52. Jasper Hopkins, *Nicholas of Cusa on Learned Ignorance: A Translation and an Appraisal of De Docta Ignorantia* (Minneapolis: The Arthur J. Banning Press, 1981), p. 1.

53. Watts, p. 48. In a letter, Cusa explains his debt to Dionysius the Areopagite, often called Pseudo-Dionysius:

> Denys, presque partout, a transmis la théologie sous forme disjonctive ... il saute par-dessus la disjonction jusqu'à la copulation et coïncidence, c'est-à-dire jusqu'à une union parfaitement simple, qui n'est point latérale mais va directement au-dessus de toute ablation et de toute position, là où l'ablation coïncide avec la position, et la négation avec l'affirmation; et telle est la plus secrète théologie, à laquelle aucun des philosophes n'est parvenu ni ne peut parvenir s'il se tient au principe commun de toute philosophie, selon lequel deux contradictoires ne coïncident point. C'est pourquoi il est nécessaire que celui qui théologise sur le mode mystique, au-dessus de toute raison et de toute intelligence, allant jusqu'à s'abandonner lui-même, pénètre dans la ténèbre; et il découvrira comment ce que la raison juge impossible, savoir: qu'une chose tout ensemble soit et ne soit pas, est la nécessité même, disons plus: si n'apparaissait pareille ténèbre et densité d'impossibilité,

la suprème nécessité n'éxisterait point, laquelle n'est pas en contradiction avec cette impossibilité, car l'impossibilité est la véritable nécessité même. (Nicolas de Cues, *Lettres aux moines de Tégernsée sur la docte ignorante* [Paris: O.E.I.L., 1985], p.27.)

[my translation]: Dionysius, almost everywhere, conveyed a disjunctive theology.... he leaps through the disjunction to copulation and coincidence, that is to say, to a perfectly simple union, which is not at all indirect but goes directly above all absence and all presence, to where absence coincides with presence, and negation with affirmation. Such is the most secret theology, to which none of the philosophers has arrived nor could arrive if he maintains the accepted principle of all philosophy, according to which two contradictories do not coincide at all. That is why it is necessary that anyone who does theology in the mystical mode, above all reason and all intelligence, going so far as to abandon himself, must penetrate into the shadows; and he will discover how to know that what reason judges impossible—that a thing can be and not be at the same time—is necessity itself. Let us say more: if such shadowy and thick impossibility does not appear, then the supreme necessity does not exist at all, which is not in contradiction with that impossibility, because the impossibility is the true necessity itself.

Thus Cusanus raises the *coincidentia oppositorum* to a test of theology itself: if a piece of theology does not encounter irrational opposites, then it has not attained its true height.

54. Jasper Hopkins, *A Concise Introduction to the Philosophy of Nicholas of Cusa*. 3rd ed. (Minneapolis: The Arthur J. Banning Press, 1986), p. 13.

55. Cassirer, p. 45.

56. Cassirer, p. 14.

57. "There Is No Natural Religion" [a], E2; and "There is No Natural Religion" [b], E2–3.

58. Hazard Adams, *Philosophy of the Literary Symbolic* (Tallahassee: Univ. Presses of Florida, 1983), pp. 27, 114.

59. Adams, *Philosophy*, p. 114.

60. Adams, *Philosophy*, p. 109.

61. Watts, pp. 229–30.

62. Carl Jung, *Answer to Job*, trans. R.F.C. Hull (1958; rpt. Princeton UP, 1973), p. x.

63. Jung, *Answer*, p. 7.

64. Jung, *Answer*, pp. 91–92.

65. Jung, *Answer*, p. 98.

66. Jung, *Answer*, p. 45.

Chapter 2

"Whatever is, is right": Learned Ignorance and the Coincidence of Opposites in *An Essay on Man*

Alexander Pope concludes his *Essay on Man* by asserting that

> WHATEVER IS, IS RIGHT;
> That REASON, PASSION, answer one great aim;
> That true SELF-LOVE and SOCIAL are the same;
> That VIRTUE only makes our Bliss below;
> And all our Knowledge is, OURSELVES TO KNOW.
> (IV, 394–398)[1]

The triumphant tone of these proclamations may induce the reader to forget that they proudly oppose man's common knowledge. The great poet of common sense seems intent on blasting common sense. The first assertion dismisses the existence of evil in the world; the second unites a pair of exact opposites as does the third; the fourth denies almost all man's strivings after happiness; the last implicitly denies the existence of God. This discrepancy between the tone of certainty and the content of contradiction is but the final opposition in a poem full antitheses.[2]

The entire argument of *Essay on Man* and many of the particulars of Pope's thoughts in it seem to follow those of Nicolas Cusanus, whose ideas I have discussed in Chapter 1. This passage from the end of the poem makes foolish most of man's wisdom, as it, like the entire poem, encourages the reader to let go of his insistence on reason, on the Law of Non-contradiction, in order to achieve a higher vision, or at least a higher acceptance of the universe of God. Like Cusanus, Pope uses techniques of learned ignorance to present his message.

As shown in this passage, the task of the poet is to inform us that we have made the wrong separations: ourselves from God, ourselves from each other, and even ourselves from ourselves. Once he makes us aware of the separations, he then reconciles them in his poem and urges us to do likewise in our lives. In trying to teach the reader this lesson, the poem moves from wandering confusion to blissful certainty, which resolves the very oppositions which had produced that uncertainty. Apparent separations are resolved into intermingled parts of the whole.

The most important opposition in the poem, around which all the others focus, is the infinite gulf between God and man. That infinite gulf was also the major concern of Cusanus, who proposed the solution of Christ, the coincidence of the opposites of God and man. Although Pope was also a staunch Catholic, he does not explicitly introduce Christ into his poem. But in a way similar to that of Cusanus, Pope insists that since man's finite limitations can never grasp the infinity of God, he must go beyond ordinary human logic in order to think about God at all. Cusanus's formulation fits Pope's poem exactly:

> In every enquiry men judge of the uncertain by comparing it with an object presupposed certain, and their judgment is always approximative; every enquiry is, therefore, comparative and uses the method of analogy.... the infinite as infinite is unknown, since it is away and above all comparison.[3]

No matter how much man learns, his learning is but ignorance: the more he knows about himself and about God, the more he knows that he does not know. He does not find himself any closer to God than he was before; he finds himself instead in a state of learned ignorance. This oxymoron foreshadows the larger coincidence of opposites that learned ignorance leads to. If God, who is beyond our direct knowledge, is infinite in time and space, then he includes everything in time and space. He is A and non-A at the same time and to the same degree. Therefore ordinary Aristotelian logic, which does not allow such contradictions, fails the thinker who wants to investigate the infinite. Such a thinker needs a logic which allows for the coincidence of opposites, for the co-existence of A and non-A in defiance of the rules of ordinary logic.

How can God and man be united when they are separated by an infinite gap? The answer to that question, worked out through the course of the poem, is encapsulated in the epigrams, quoted above, which end the poem. A summary of the meanings of the last lines of the poem will prepare for a more detailed examination of the poem as a whole.

In the chiasmus, "Whatever is, is right," the world of finite man and the universe of infinite God are the terms. The first two words posit the existence of certain finite facts which can be verified by observation; the last two words reverse the terms and see them from an infinite perspective, which cannot be verified by man's direct observation, but which must be accepted on faith. "Whatever" exists in the finite world is "right" from the perspective of the infinite world.[4]

As I point out in my introduction, Leopold Damrosch in *Symbol and Truth in Blake's Myth* says that a true coincidence of opposites can be achieved only by someone who fundamentally accepts things as they are: "A true acceptance of opposites demands a skeptical temperament and a commitment finally to things as they are."[5] This kind of complacency, so attacked by the Romantics, is the foundation of Pope's poem. He allows his elitist conception of society to infect his picture of the universe. In his satires he will excoriate the abuses he sees; here he seems rather to excuse them.

Elizabeth Tebeaux explains how Pope's ideas here fit into the tradition of Pyrrhonism.[6] In that form of skepticism, rational arguments are rejected because all arguments generate equally valid counter-arguments. Many Catholic Pyrrhonists, like Pope, used this principle to suspend judgment and to allow themselves to accept the teachings of the Church, since their reason was incapable of answering ultimate questions on its own.

At the beginning the poet's easygoing manner, that of a country gentleman strolling around his estate, complacently surveys the dangerous mysteries which lurk behind the foliage. He will

> Expatiate free o'er all this scene of Man,
> A mighty maze! but not without a plan;

A Wild, where weeds and flow'rs promiscuous shoot,
Or Garden, tempting with forbidden fruit.
(I, 5–8)

Throughout the poem Pope seems to stand complacent before the horrors of existence. Where the environment is wild, there beautiful and ugly sex and violence burst forth; where it is tamed, there the temptation of the fruit of the tree of the knowledge of good and evil threatens man's peace. Douglas Canfield sees the theme of the Fall worked in throughout the poem. Although the poem explicitly seems to deny the Fall, frequent images of rising and falling and frequent use of the word "fall" imply a theological stance:

> Pope insists again and again that the entire world, including man, could never have existed in any other than its present state of imperfection.[7]

In fact, in Canfield's interpretation, Pope starts to sound quite like Blake:

> Pope reinterprets the Fall as a metaphor for something which occurs not once but perennially, synchronically, in man's continual fall through 'reas'ning pride.'[8]

False reason opposes the mysterious *coincidentia oppositorum* of God, which includes all oppositions, including garden and wild.

The contrast of the garden and the wild contains one of the basic oppositions of the poem: the uncontrolled passion of self-love, which can destroy, versus the overweening pride of reason, which can presume to usurp God's knowledge. Only God's knowledge can discern the plan that informs the apparent maze;[9] man's speculations toward knowledge of the nature of garden or wild must begin with his own experience, as they must in Cusanus:

Say first, of God above, or Man below,
What can we reason, but from what we know?
(I, 17–18)

Reasoning from what we know can lead us to some knowledge of God.

Cusanus abstracts from mathematical knowledge in order to speculate about God after making a statement very similar to Pope's:

> it is only by way of postulates and things certain that we can arrive at the unknown.... but the more we abstract from sensible conditions, the more certain and solid our knowledge is.[10]

Instead of presenting us with various analogies from our world which correspond to the higher world, as Cusanus does, Pope prefers to present us with puzzles and paradoxes as a way of contradicting ordinary sense and raising us to a vision of the infinite. Man's knowledge can not rise above his own sphere, yet he can be sure that there is a grand plan through his faith. Carefully concealing any explicit consideration of God's revelation to man and to the incarnation of Christ, Pope limits his poem to the realm of man. His Christianity is kept implicit throughout.

The distance between the realm of man and the realm of God remains absolute: "'Tis but a part we see, and not a whole" (l. 60). As long as we are living, we can never hope to see the whole. Pope's emphasis on our limited knowledge places him in the skeptical tradition:

> Pope's scepticism leads him to place truth in the reality of the spiritual, the One, Whose presence and order pervade all things.... Like Nicholas of Cusa, Pope can still be very much a sceptic and find consolation in the Christian Neo-Platonic notion that man is a future creature who must die to be able to apprehend and comprehend the highest level of cognition.[11]

Any attempt to judge God by human standards foolishly falls under its own futility. Like Cusanus's learned ignorance, man's knowledge must not succumb to hubristic attitudes:

> Go, wiser thou! and in thy scale of sense
> Weigh thy Opinion against Providence;
> Call Imperfection what thou fancy'st such,
> Say, here he gives too little, there too much;
> Destroy all creatures for thy sport or gust,
> Yet cry, If Man's unhappy, God's unjust;
> If Man alone ingross not Heav'n's high care,
> Alone made perfect here, immortal there:

Snatch from his hand the balance and the rod,
Re-judge his justice, be the GOD of GOD!
(I, 113–122)

The handling of oppositions in this passage of imagined disturbance of the Great Chain of Being typifies Pope's method of confounding man's finite reason on its own terms, thus forcing man to acknowledge, if not actually understand, the need to yield to the infinite knowledge of God. The part cannot be wiser than the whole. The balanced phrases ironically echo man's attempts to upset the balance of God's universe. In line 116 "here" opposes "there"; "too little" opposes "too much." The balanced oppositions highlight the fact that such an opinion accuses God of not distributing his gifts in a balanced way, and thus shows man's refusal to acknowledge the ignorant aspect of learned ignorance. Line 113, the only line in this quotation that does not contain a balanced opposition within its limited ten syllables, scorns the puny scales of man's limited perspective in relation to the grand balance of God. Trying to weigh his tiny opinion against God's unlimited providence, the presumptuous man dares to contemplate usurping God's distribution of happiness (the balance) and therefore tries to usurp God's authority of rulership (the rod). In line 114, "thy Opinion" opposes "Providence," repeating the terms of the primary contrast between man and God. However, when in line 118 "unhappy" is balanced by "unjust," these two terms do not contradict each other; instead, their almost synonymous parallelism provides a variation on the series of antonymous parallelisms. The hypothesis of the proposition—"If Man's unhappy"—is true, certainly, but if the conclusion is true—"then God's unjust"—then man has presumed that he knows more than God knows; he has, in fact, made himself into a God over the real God.

If man is God, then God is not God; the infinite being does not exist, and man's presumptuous reason has destroyed the structure of the whole universe. However, the danger here is no more real than it is later when Pope pictures the whole universe careening out of control (I, 241–258). Rather like the story of the Tower of Babel, the device heightens the impossibility of man's attaining any such heights as he pridefully may attempt to scan. The near repetition of the word "justice" and the exact repetition of the word "God," repetitions balanced in the two hemistiches of line

122, reassure us of the security of the grand scheme, for the only judge of justice is God and the only God of God is God. This certainty of the entire scheme anticipates Blake's safety net—the impossibility of falling past certain merciful limits—that I discuss later.

It is crucial for Pope to make sure that the distinctions that he wishes to maintain are not lost in his coincidences of opposites, as later when he says that black and white still exist even though they often blend. Any theory of the coincidence of opposites first puts heavy emphasis on the opposites themselves and on their irreconcilable nature, the better to perform the miracle of coincidence or interpenetrability.

But if this God cannot be grasped by man, if He is an infinite distance away, then what is His relationship to the finite world? If He simply created the world and then stepped away, we don't need Him any more. On the other hand, if He is in everything, then there is no separation between God and the world. Pope accepts neither of these extremes. Instead he leans toward a neo-Platonic explanation which combines God's transcendence and His immanence:

> All are but parts of one stupendous whole,
> Whose body, Nature is, and God the soul;
> That, chang'd thro' all, and yet in all the same,
> Great in the earth, as in th'aethereal frame,
> Warms in the sun, refreshes in the breeze,
> Glows in the stars, and blossoms in the trees,
> Lives thro' all life, extends thro' all extent,
> Spreads undivided, operates unspent,
> Breathes in our soul, informs our mortal part,
> As full, as perfect, in a hair as heart;
> As full, as perfect, in vile Man that mourns,
> As the rapt Seraph that adores and burns;
> To him no high, no low, no great, no small,
> He fills, he bounds, connects, and equals all.
> (I, 267–280)

As Cusanus explains of the neo-Platonists:

> They were of opinion that all movement came from this Soul of the World which, they said, was entire in the whole and entire in each part of the world though its influ-

> ence is not the same in different parts; just as with the rational soul, which is entire in the whole and in each part of man, and yet its activity in the hairs and the heart is not the same.[12]

The whole equals more than the sum of its parts because God as the soul gives it life, just as man is more than the sum of his parts because he possesses a soul. The chiasmus of line 269, the terms of which are "chang'd-all" and "all-same," embodies the paradox of simultaneous unity and multiplicity.

The other chiasmus in this long passage of comparisons and oppositions occurs in the statement of the essential metaphor: "body:Nature::God:soul." (l. 268) The two chiasmi both illustrate that the infinite world reverses the terms of the finite world, even while being analogous to it. Lines 270–272 show a physical world fully infused with the presence of God. (In line 271, the ambiguous verb forms even make it sound as if God himself enjoys the sun and the wind.) The virtual tautologies of line 273 imply that without God as the soul life does not exist. The anaphora of lines 276–277 echoes the content: God is no less full and perfect in hair than in heart, in man than in seraph. The two lines of ten monosyllables each (lines 269, 279) show the equalizing effect which God exerts on the creations (and he is equal to all of it). In line 280 the four verbs sum up much of the poem: he fills all by virtue of his immanence; he bounds all by virtue of his position as creator; he connects all through the power of his love; and he equals all because he is fully in every thing.

When A.O. Lovejoy discusses this concept in many writers, his insistence on unit-ideas and Aristotelian logic leads him to comment:

> He [God] was the Idea of the Good, but he was also the Idea of Goodness; and though the second attribute was nominally deduced dialectically from the first, no two notions could be more antithetic. The one was an apotheosis of unity, self-sufficiency, and quietude, the other of diversity, self-transcendence, and fecundity.[13]

Lovejoy further complains of

> the permissibility and even necessity of contradicting oneself when one spoke of God.... It might appear easy to affirm of the divine nature what to us must seem incompatible

> metaphysical predicates; but it was impossible to reconcile in human practice.[14]

Lovejoy thus reveals the difficulty of accommodating the elusive coincidence of opposites that Pope is asserting here.

God's simultaneous immanence and transcendence make it absolutely clear that each aspect of finite life finds its correspondence and even its *raison d'être* in the infinite life; the perspective of any part is necessarily only a part of the perspective of the whole:

> All Nature is but Art, unknown to thee;
> All Chance, Direction, which thou canst not see;
> All Discord, Harmony, not understood;
> All partial Evil, universal Good:
> And, spite of Pride, in erring Reason's spite,
> One truth is clear, 'Whatever IS, is RIGHT.'
> (I, 294)

The poet insists that his reader submit to an infinite perspective which man cannot know, see, or understand as fully as God can. The four sets of apparent opposites hammer home the message: finite Nature equals infinite Art; finite Chance equals infinite Direction; finite Discord equals infinite Harmony; finite Evil equals infinite Good; finite actuality equals infinite correctness. Any refusal to accept these paradoxical formulas brands the recalcitrant reader as a presumptuous, prideful rebel who desires to upset the whole universe, as a weak creature who has let his reason blind him to his own weakness and who has set himself up as the judge of God, in an attempt to disturb the Great Chain of Being by presuming that learning is not ignorant and that man is equal to God in the wrong way.

Immediately after this passage, having peremptorally disposed of any wrong notions of man's relationship to God, Pope begins Epistle II with a portrait of man himself. Since we know that God eludes our analysis, perhaps we can do better looking at ourselves, by beginning with what we can know. However, even within our own realm, our own qualities and desires baffle and confuse us:

> Know then thyself, presume not God to scan;
> The proper study of Mankind is Man.

Plac'd on this isthmus of a middle state,
A being darkly wise, and rudely great:
With too much knowledge for the Sceptic side,
With too much weakness for the Stoic's pride,
He hangs between; in doubt to act, or rest,
In doubt to deem himself a God, or Beast;
In doubt his Mind or Body to prefer,
Born but to die, and reas'ning but to err;
Alike in ignorance, his reason such,
Whether he thinks too little, or too much:
Chaos of Thought and Passion, all confus'd;
Still by himself abus'd, or disabus'd;
Created half to rise, and half to fall;
Great lord of all things, yet a prey to all;
Sole judge of Truth, in endless Error hurl'd:
The glory, jest, and riddle of the world!
(II, 1–18)

Once man has accepted the contradictions between his wisdom and God's, he must now accept the contradictions within himself. At first reading, the passage seems simply to delineate, in the most cynical manner possible, the hopeless middle state in which man finds himself, but upon closer examination, it reveals the seeds of hope that will find fruition in the progress of the poem, after long and difficult struggle.

Included in the harshly phrased realities are the equally real optimistic possibilities, which will build back up to reunite the man and God who were so definitely separated in Epistle I. First of all, an isthmus, however gloomy, is not as gloomy as an island. An isthmus, although threatened on two sides by oceans, is also connected on its two ends to the mainlands, in this case God and Nature. Furthermore, man is wise and great, albeit his wisdom sees through a glass darkly and his greatness is primitive compared to God's. Because he knows of his own darkness and rudeness, he may want to reject reason altogether and become a sceptic. However, he has too much knowledge to do that. But if he puts too much faith in his knowledge, then he may take on the pride of a stoic, who relies exclusively on reason. His own inescapable weaknesses, however, prevent that attitude. Pope thus refuses to allow man the comfort of embracing either extreme or the comfort of ignoring either extreme. In fact, in Pope's formulation, the limits of human knowledge seem positively good instead of to be resented.

Man cannot change the position in which God has placed him, but he can choose what to do about it. The oppositions have been carefully chosen and balanced, and are absolutely true, unmistakably admitting the existence of two extremes in each case. But man's thinking too much or too little leads him into various errors, in which he separates those things which should not be separated. In Epistle I it is very important to make the correct separations between oppositions; in Epistle II it is very important to put together things which should not be separated.

The poet allows foolish man to pose false separations: action or rest, God or beast, mind or body. Each of these three pairs is preceded by the anaphoric "in doubt," the opposite of an implied "in faith." The faith which led to an acceptance of the coincidences of opposites in the universal scheme at the end of Epistle I now has turned to doubt, which refuses the oppositions within the individual, and tries to make false choices. Instead of admitting his middle state, man tries to make up his mind about which of the two extremes he really belongs to. He should certainly lean toward the first of each of the pairs; he should prefer the first terms in their proper hierarchy, but he should not deny the lower terms. And if he allows himself to think too curiously on his condition, he will become too passive:

> Fix'd like a plant on his peculiar spot,
> To draw nutrition, propagate, and rot.
> (II, 63–64)

Of course on the other extreme, thinking too little and taking too much action, like a physical beast, will make him

> meteor-like, flame lawless thro' the void,
> Destroying others, by himself destroy'd.
> (II, 65–66)

If man allows himself to fall into these false distinctions and separations, then he can become a chaos of thought and passion, all confused. This condition echoes the condition before the creation, before God said, "Let there be light." Once God, in the original act of Creation, has created the light and separated it from the darkness, it is up to man to maintain or re-create that balance or alternation within himself. Later in the Epistle he is instructed how to do so:

> This light and darkness in our chaos join'd,
> What shall divide? The God within the mind.
> (II, 203–204)

The power of God within the mind must re-establish order within the mind. Thus Pope implies a dynamic, continual creation, analogous to that of Blake. He also anticipates Coleridge's definition of the imagination in which man's primary and secondary imaginations are echoes of the great "I AM" of God. Man does not create the order, but must re-create within himself the order which God has already created. Man's power is analogous to God's, which was so definitely and vehemently separated from man's in Epistle I.

In between the chaos at the beginning of Epistle II (1–18) and the seemingly obvious and simple solution near the end of the epistle (203–204) the poet explains the understanding necessary to re-create God's order. First he traces the inevitable course of a proud reason which tries to reach God by its own power. If it tries to rise in the wrong way, then in effect it tries to usurp the place of God. Trying to escape the lowness and confusion of the human condition, man may proudly try to mount above his condition, but he then finds that too much height can lead to sudden depth:

> Go, wond'rous creature! mount where Science guides,
> . . .
> Go, soar with Plato to th' empyreal sphere,
> . . .
> And quitting sense call imitating God;
> . . .
> Go, teach Eternal Wisdom how to rule—
> Then drop into thyself, and be a fool!
> (II, 19, 23, 26, 29–30)

The extremes, however confusing and painful, cannot be simply escaped. And straining to go too high leads to the diametrical opposite of depth. In either case, man is trying to explore the self; this poem insists that there is a right way and a wrong way to do so. The mind must learn to exist in a state of dynamic tension combining contraries, a simultaneous separation between man and God with a recognition of the combinations of apparent opposites. Earl Wasserman, writing of the *concordia discors* in "Windsor Forest," succinctly generalizes Pope's point:

> when an appetite is not brought into concordant clash with a contrary force it paradoxically both grows to its own excess and in this act destroys itself by becoming its own opposite.[15]

Or, to put it in words which Pope himself had read:

> He, who acts in a conformity to the nature of things, carries on the system of God, and cooperates with him: and surely to put the system of divine wisdom in execution, and to cooperate with the creator is honor enough for the creature. Thus we may attain to the perfection of our nature, and, by pretending to no more, we may do it real honor: whereas, by assuming that we imitate God, we give the strongest proof of the imperfection of our nature, whilst we neglect the real, and aspire vainly at a mock honor.[16]

A man who climbs with his proud reason thinks he is becoming more self-sufficient. But self-sufficiency is possible for God only, not for man. If man presumes to becomes God, he presumes to become complete unto himself; the result, however, is not ultimate self-knowledge, but ultimate self-deception, self-abuse. Man alone does not rise, but instead drops into himself. Instead of knowing himself by trying to see what he is, he falls into the very lowness that he was trying to escape.

The proper condition of reason is prescribed in II, 43–46.

> Trace Science then, with Modesty thy guide;
> First strip off all her equipage of Pride,
> Deduct what is but Vanity, or Dress,
> Or Learning's Luxury, or Idleness.

Reason must be spartanly stripped down to its essentials, accompanied by modesty, not pride.

Once Pope has shown the wrong and right uses of reason, he contrasts it with self-love. Although the impulses of the two forces are contrary to each other, Pope explicitly refuses to exalt one over the other:

> Two Principles in human nature reign;

Self-love, to urge, and Reason, to restrain;
Nor this a good, nor that a bad we call,
Each works its end, to move or govern all:
And to their proper operation still,
Ascribe all Good; to their improper, Ill.
(II, 53–58)

Pope anticipates Blake in his refusal to separate good and bad absolutely. Rather the separation is in the use of the forces. Passion and reason are thus in a reciprocal relationship, each necessary to the other.

Three metaphors for the relationship between passion and reason explain the complex balancing that man must maintain within himself. These images progress from the mechanical through the natural to the creative. A mechanical metaphor makes self-love and reason into parts of a watch:

Self-love, the spring of motion, acts the soul;
Reason's comparing balance rules the whole.
(II, 59-60)

Another metaphor, chosen from nature, has the two principles performing complementary functions, but in this one reason exerts no force of its own:

The rising tempest puts in act the soul,
Parts it may ravage, but preserves the whole.
On life's vast ocean diversely we sail,
Reason the card, but Passion is the gale;
Nor God alone in the still calm we find,
He mounts the storm, and walks upon the wind.
(II, 105–110)

Whether the word "card" is interpreted to mean map or compass, in either case it still fills only an advisory function. In another metaphor, however, reason becomes a vigorous artist:

Passions, like Elements, tho' born to fight,
Yet, mix'd and soften'd, in [God's] work unite:
Suffice that Reason keep to Nature's road,
Subject, compound them, follow her and God.
Love, Hope, and Joy, fair pleasure's smiling train,
Hate, Fear, and Grief, the family of pain;
These mix'd with art, and to due bounds confin'd,

Make and maintain the balance of the mind:
The lights and shades, whose well accorded strife
Gives all the strength and colour of our life.
(II, 111–122)

Here the balance to be maintained is not between reason on the one hand and passion on the other, but between two sets of passions. Like the dark and light paints on an artist's palette, the passions must be mixed and ordered by reason, which confines the passions to their proper bounds just as God confined the elements at the creation.[17]

The power of the infinite dwells in men both in their reason ("the God within the mind") and in their passion ("like Aaron's serpent"), and so reason and passion can work together to help man find happiness. Self-love, the desire for sexual gratification and for safety, founds families and societies, and so unites individual and social goals. The bliss that men seek on this earth comes only from acknowledging the rightness of God's plan, incorporating the power of God inside themselves in their reason and their passion, loving those outside themselves as much as they love the souls within themselves, and realizing that only virtue allows men to find this bliss. Paradoxically, to find this proper relationship to God, self, and others, man must admit that he can know nothing but himself; he must accept learned ignorance.

Having shown us the extreme of overweening reason and the proper balance of reason and passion, Pope then proceeds to show us the extreme of passion in any individual, which may be termed one's "ruling passion":

... one master Passion in the breast,
Like Aaron's serpent, swallows up the rest.
(II,131–132)

The image of Aaron's serpent swallowing up the serpents of the magicians of Pharaoh implies that one's ruling passion is a sign of God's power, that He sends it to work his ends. In similar fashion one's ruling passion delivers a person from the confusion and chaos which beset him because it allows him a chance to organize his life around some consistent principle.

Bertrand Goldgar points out that Pope makes relatively new use of a complex body of ideas about the passions. Goldgar even goes so far as to assert that

> Pope elevates the notion of a ruling passion into the principle which makes the ultimate reconciliation between the elements of apparent discord in man's nature and the contrasting attitudes which he has held in balance throughout the Epistle.[18]

Goldgar's assertion seems to over-ride the savage destructiveness of the ruling passion (II, 133–138) and the continuing necessity for reason (II, 197). As a force analogous to instinct in animals, it is in a sense a direct manifestation of the power of God within man. However, the ruling passion not only fails creatures with life, but also destroys them:

> As Man, perhaps, the moment of his breath,
> Receives the lurking principle of death;
> The young disease, that must subdue at length,
> Grows with his growth, and strengthens with his strength:
> So, cast and mingled with his very frame,
> The Mind's disease, its ruling Passion came.
>
> (II, 133–138)

The implication is that man begins to die as soon as he is born and even more than that, the cause of his death is the same force which is the cause of his active life. Here the relationship between reason and passion is a savage one; like a parasite, the ruling passion parallels the growth of the mind and constantly threatens to destroy it. Because the principle of death does indeed destroy every person, the image implies that the ruling passion always destroys the mind.

Extreme ease (as in the earlier dichotomy between body and mind, rest and action), would turn man into a vegetable, and so some dis-ease is necessary to animate him, but surely the poet expresses deep melancholy here (perhaps colored by his own view of his life as a long disease) in the admission that life is inextricably entwined with death. The final hemistiches of lines 133–134 balance the "moment" of breath against the "principle" of death. The former is fleeting, finite; the latter, eternal, infinite.

Although the ruling passion unites and directs the efforts of men, it violates a very important principle which is the direct cause of its destructiveness. Instead of vital humors flowing to the whole and thus creating the proper balance in the body and soul, they flow only to the ruling passion, thus creating an imbalance which leads to destruction (II, 139–150). Just as proud reason can repeat the sin of Eve in desiring to know all that God knows, so the ruling passion can usurp the wholeness of man by trying to control all.

Each individual can, however, attempt to imitate the grand balance of God's whole universe by following the better aspect of his ruling passion:

> Th'Eternal Art educing good from ill,
> Grafts on this Passion our best principle:
> 'Tis thus the Mercury of Man is fix'd,
> Strong grows the Virtue with his nature mix'd;
> The dross cements what else were too refin'd,
> And in one interest body acts with mind.
> As fruits ungrateful to the planter's care
> On savage stocks inserted learn to bear;
> The surest Virtues thus from Passions shoot,
> Wild Nature's vigor working at the root.
> What crops of wit and honesty appear
> From spleen, from obstinacy, hate, or fear!
> See anger, zeal and fortitude supply;
> Ev'n av'rice, prudence; sloth, philosophy;
> Lust, thro' some certain strainers well refin'd,
> Is gentle love, and charms all womankind:
> Envy, to which th'ignoble mind's a slave,
> Is emulation in the learn'd or brave:
> Nor Virtue, male or female, can we name,
> But what will grow on Pride, or grow on Shame.
> Thus Nature gives us (let it check our pride)
> The virtue nearest to our vice ally'd;
> Reason the byass turns to good from ill.
>
> (II, 175–197)

And so the wild at the beginning of the poem becomes one with the garden. Using the knowledge he has now gained about God's relation to the world and about the proper relation of reason and passion, man can use the God within his mind to re-create the order which belongs there.

Immediately after the phrase "The God within the mind," Pope draws an analogy between the function of extremes within the individual and the function of extremes within the world:

Extremes in Nature equal ends produce,
In Man they join to some mysterious use;
Tho' each by turns the other's bound invade,
As, in some well-wrought picture, light and shade,
And oft so mix, the diff'rence is too nice
Where ends the Virtue or begins the Vice.
  Fools! who from hence into the notion fall,
That Vice or Virtue there is none at all.
If white and black blend, soften, and unite
A thousand ways, is there no black or white?
(II, 205–214)

On one level, this warning reminds us that no matter how necessary it is to unite opposites, we must always remember that they remain opposites still. If they lose their own characters, then we lose the dynamic balance. But a much more important message dwells in these lines. A careful distinction must be made between the opposites of reason and passion, both of which are beneficial if used correctly, and the opposites of vice and virtue, one of which is to be shunned. This distinction anticipates the distinctions between Contrary and Negation in Blake and between contrary and opposite in Coleridge. Although it is impossible to extirpate vice entirely (II, 231–232), we should not try to balance it with virtue as we should try to balance reason and passion. Vice does not work with virtue, but cancels it out.

But even if individual men fail to balance their reason and their passion and choose vice over virtue, still the infinite wholeness of God maintains the integrity of the universe:

Each individual seeks a sev'ral goal;
But HEAV'N's great view is One, and that the Whole:
That counter-works each folly and caprice;
That disappoints th' effect of ev'ry vice:
...
Heav'n forming each on other to depend,
A master, or a servant, or a friend,
Bids each on other for assistance call,
'Till one Man's weakness grows the strength of all.
(II, 237–240, 249–252)

See! and confess, one comfort still must rise,
'Tis this, Tho' Man's a fool, yet GOD IS WISE.

(II, 293–294)

After humbling man in Epistle I and humiliating him in Epistle II, the poet begins to reassure him in Epistle III by showing him the cycle of Nature. Pope has broken down man's preconceptions in order to build up a better view of the universe. Man does not need to trust to himself alone, indeed must not do so, but instead must see himself as part of the whole, where strength and weakness, life and death sustain each other:

> Look round our World; behold the chain of Love
> Combining all below and all above.
> ...
> See dying vegetables life sustain,
> See life dissolving vegetate again
> ...
> Nothing is foreign: Parts relate to whole;
> ...
> All serv'd, all serving! nothing stands alone;
> The chain holds on, and where it ends, unknown.
>
> (III, 7–26)

A view of the supposedly lowly beast reveals to proud man that animals do not have to worry about combining reason and passion, for there is no distinction between their instinct and their reason. Although Pope does not quite say so, this point is reminiscent of Cusanus's point that reason in men is equivalent to instinct in beasts:

> See then the acting and comparing pow'rs
> One in their nature, which are two in ours,
> And Reason raise o'er Instinct as you can,
> In this 'tis God directs, in that 'tis Man.
>
> (III, 95–98)

Understanding this principle can help men to understand that the self-love that they consider to be part of their lower nature can be used for the good of all. Just as the ruling passion can be transmuted into virtue instead of vice, so the passionate desire for sexual gratification is the foundation of the family and therefore of society:

> Each loves itself, but not itself alone,
> Each sex desires alike, 'till two are one.

Nor ends the pleasure with the fierce embrace;
They love themselves, a third time, in their race.
(III, 121–124)

Another aspect of self-love leads to another aspect of the establishment of society, which fully combines the opposites of self and society:

All join to guard what each desires to gain.
Forc'd into virtue thus by Self-defence,
...
Self-love forsook the path it first pursu'd,
And found the private in the public good.
(III, 278–282)

Thus the desire for sex and the desire for safety, both springing from self-love, benefit the society as a whole.

Thus God and Nature link'd the gen'ral frame,
And bade Self-love and Social be the same.
(II, 317–318)

Just as each man's mind must re-create in itself the order which God first created, so mankind as a whole must re-create in his environment the Eden which God first gave him by re-uniting wild with garden and reason with passion.

The platitudes in the beginning of Epistle IV seem a far-too-simple resolution of the complicated concepts which the poet has fronted in the earlier epistles:

Take Nature's path, and mad Opinion's leave,
All states can reach it, and all heads conceive;
Obvious her goods, in no extreme they dwell,
There needs but thinking right, and meaning well.
(IV, 29–32)

The simplicity which the poet asks us to believe in has been hard won out of the complexities, just as the simple realization of the unity of infinity must be hard won out of the multiplicity of direct experience. The man who finds happiness is he who

Pursues that Chain which links th'immense design,
Joins heav'n and earth, and mortal and divine;
Sees, that no being any bliss can know,
But touches some above, and some below;

> Learns, from this union of the rising Whole,
> The first, last purpose of the human soul;
> And knows where Faith, Law, Morals, all began,
> All end, in LOVE of GOD, and LOVE of MAN.
> (IV, 333–340)

This linking of the finite and the infinite is the beginning and end of the poem. A staunch Christian, Pope has written an entire poem about the coincidence of finite and infinite without once mentioning Jesus Christ, the prime example of such a coincidence of opposites. But the voice of Christ enters before the poem concludes:

> Self-love thus push'd to social, to divine,
> Gives thee to make thy neighbour's blessing thine.
> Is this too little for the boundless heart?
> Extend it, let thy enemies have part.
> (IV, 353–356)

This passage echoes a portion of Christ's Sermon on the Mount:

> Ye have heard that it was said, Thou shalt love thy neighbor, and hate thine enemy; but I say unto you, Love your enemies, and pray for them that persecute you; that ye may be sons of your Father who is in heaven: for he maketh his sun to rise on the evil and the good, and sendeth rain on the just and the unjust. For if ye love them that love you, what reward have ye? do not even the publicans the same? And if ye salute your brethren only, what do ye more than others? do not even the Gentiles the same? Ye therefore shall be perfect, as your heavenly Father is perfect.
> (Matthew 5:44–48)

Not only do these words provide much of the material of *An Essay on Man,* including the most difficult coincidence of opposites—love (instead of hate) your enemies (not only your friends)—they also promise infinite perfection for man. Pope has gone to great pains to convince his reader that he must accept his humanity and not presume to be God, and yet here he invokes words of Christ which explicitly promise that greatest coincidence of opposites: man can be as perfect as God is, not perfect in his nature, but perfect in his actions.

The paradox can be solved only be refusing to dissolve the paradox. Man's condition must be accepted on faith. If God can be immanent and transcendent at the same time, and if all the other coincidences of opposites in the Essay can be true, as Pope has asserted that they are, then such things can exist only if man refuses to be bound by finite logic even while speaking in finite words. And this is precisely the tour de force which Pope has performed. Using not only limited words but the limited form of the heroic couplet, and claiming to limit his inquiry to the condition of man, the poet brings the reader to an understanding of the mysterious unity of finite and infinite in God, in Christ, in the poem, and in man. Analogous to the Christ who unites God and man, the poem becomes the mediator which unites poet and word and leads us into the logic of the infinite by taking a specific limited form.

The poem contains so many antitheses because it takes on the form of its subject: learned ignorance and the coincidence of opposites which necessarily follows. The mind can approximate understanding of the infinite by intentionally looking at coincidences of opposites until the mind performs a leap into a different dimension. Similarly, Pope's overwhelming variations on antitheses and opposition, both in content and in style, force the reader either to dismiss the poem as illogical confusion or to accept it as super-logical truth. Both come together in the injunction to love enemies, because knowledge has become less important than right action.

By embracing the seemingly nonsensical command to love his enemies, man reciprocates the benevolence of God:

> God loves from Whole to Parts: but human soul
> Must rise from Individual to the Whole.
> (IV, 361–362)

Thus man's love is identical to that of God, except that it is reversed, a mirror image. God's love unquestionably takes in all of creation; man, starting from his own finite position, can mirror that love by extending his natural self-love into universal benevolence, starting from self-knowledge which realizes its limitations and can overcome those limitations only by accepting them in the first place. But man can accomplish this expansion of his epistemology and his ontology only if he contradicts his natural finite concepts with infinite ones and accepts the mystery of the universe as it appears to God. And the only way that he can achieve that vision is to accept himself as part of that total universe. He cannot know God directly; he cannot even know the rest of creation. But he can partially know himself and allow his own opposites to coincide. In this sense, then, a part can contain the whole. Not only the immanence of God, but man's full knowledge of himself makes him part of the holding together of the universe. Only if he accepts the coherence of the universe can he accept himself, and only if he accepts himself can he accept the coherence of the universe.

The poem serves as a mirror in which the reader can confront the opposites within the universe and within himself, and see them duplicated, and yet contradicted and reversed. The mirror of art and the immanence of God in the form of Christ together illuminate the finite in a way which aids it to see beyond itself.

> For Wit's false mirror held up Nature's light;
> Shew'd erring Pride, WHATEVER IS, IS RIGHT;
> That REASON, PASSION, answer one great aim;
> That true SELF-LOVE and SOCIAL are the same;
> That VIRTUE only makes our Bliss below;
> And all our Knowledge is, OURSELVES TO KNOW.

Notes

1. All quotations of *An Essay on Man* are taken from the Twickenham edition, Maynard Mack, ed. (1950; rpt. New Haven: Yale University Press, 1964).

2. Martin Kallich has counted the antitheses. The number is immense, larger than in any of Pope's other poems. See "Unity and Dialectic: The Structural Role of Antitheses in Pope's *Essay on Man*," in *Papers on Language and Literature*, 1 (1965), 109–24.

3. Cusanus, *Of Learned Ignorance*, p. 7.

4. Compare John Locke's much less distressing (and much less elegant) phrase, "Whatever is, is," at the beginning of his *Essay Concerning Human Understanding*.

5. Damrosch, p. 242.

6. "Scepticism in Pope's 'Essay on Man,'" *College Literature* 10 (1983): 158–71.

7. Douglas Canfield, "The Fate of the Fall in Pope's *Essay on Man*," *The Eighteenth Century* 23 (1982), p. 142.

8. Canfield, p. 144.

9. William Empson explains how the basic meanings and ambiguities of this passage remain the same, or at least just a complicated, even in a manuscript reading, "A mighty maze of walks without a plan." *Seven Types of Ambiguity* (1930; rvsd. ed., New York: New Directions, 1947), p. 204.

10. *Of Learned Ignorance*, p. 25.

11. Tebeaux, p. 160.

12. Cusanus, *Of Learned Ignorance*, p. 99.

13. Lovejoy, pp. 82–83.

14. Lovejoy, p. 83.

15. Earl Wasserman, *The Subtler Language* (1959; rpt. Baltimore: Johns Hopkins UP, 1964), p. 63.

16. Henry St. John, Lord Viscount Bolingbroke, qtd. in Kallich, "Unity," p. 32.

17. Leo Spitzer points out that, in Latin texts dealing with the *concordia discors*, the frequency of words containing the syllable "con," meaning "with," is a visual and aural reminder that the subject is the joining of opposing things. The frequency of "con" (and "com") is of course very high in Pope's poem and in my commentary on it. *Classical and Christian Ideas of World Harmony* (Baltimore: Johns Hopkins UP, 1963), p. 149.

18. Bertrand Goldgar, "Pope's Theory of the Passions: The Background of Epistle II of the *Essay on Man*," *Philological Quarterly*, 41 (1962), p. 731.

Chapter 3
Blind to the Real Presence: Coleridge and the Tension of Opposites

As much as any writer, and certainly more than most, Samuel Taylor Coleridge actively seeks for coincidences of opposites. According to N.M. Goldsmith,

> Coleridge was thinking out thoughts which had been half conscious in Pope and a number of other men of a religious temperament since the seventeenth century, ... the feeling that a mechanistic theory of the mind which denied the validity of intuition, concentrating instead on men's capacity for deduction and analysis, was a philosophy of death.[1]

Thus, as Coleridge develops his theories of the coincidence of opposites and other theories of the way the mind works, he will be carrying on many of the ideas that Pope developed, although in most cases he tries to take them further.[2]

Unlike Pope, Coleridge gives a particularly personal slant to the problem of the coincidence of opposites. Sometimes, especially in his philosophical prose, Coleridge does present a generalized, overall discussion of the problem, but most often, especially in his poetry, he explores the issue from an intensely personal point of view. The problem tears him apart and he reveals that tearing most painfully and most despairingly.

I do not intend to review all aspects of his struggles with opposites, which have been thoroughly studied by many critics, but simply to set out some of the main lines of Coleridge's interest, leading up to one of his late poems, "Lines Suggested by the Last

Words of Berengarius," a poem that epitomizes the dead end in which Coleridge found himself when he tried to reconcile opposing forces in his poetry. My line of inquiry will show how, even though Coleridge seems largely to come to terms with the issue of the *coincidentia oppositorum* in his philosophical writings, he finds dead ends instead of solutions in his poetry. I do not belittle Coleridge. I admire his bravery as much as Richard Holmes and Thomas McFarland do in passages quoted below. Rather, his heroic struggles give us a sense of the enormous difficulty of the problem and allow us to admire Shelley's and Blake's solutions even more.

Coleridge's desire to reconcile opposites is closely connected to his desire to see wholeness whenever possible. In his essay "On Method" he defines the superior man in terms of wholeness:

> What is that which first strikes us, and strikes us at once, in a man of education, and which, among educated men, so instantly distinguishes the man of superior mind? Not the weight or novelty of his remarks; not any unusual interest of facts communicated by him.... It is the unpremeditated and evidently habitual arrangement of his words, grounded on the habit of foreseeing, in each integral part, or (more plainly) in every sentence, the whole that he then intends to communicate. However irregular and desultory his talk, there is method in the fragments.[3]

Similarly, when Coleridge contemplates writing an epic, the encyclopedic genre of wholeness, he can be satisfied with nothing less than all. While most writers of epic do feel a need to surpass the accomplishments of their predecessors,[4] Coleridge seems unable to omit anything. In a letter to Joseph Cottle in early April 1797, after agreeing with Wordsworth "that Southey writes too much at his ease," Coleridge laments Southey's reliance "too much on story and event in his poems, to the neglect of those lofty imaginings, that are peculiar to, and definitive of, the poet." Meditating by contrast on Milton, Samuel Taylor Coleridge imagines his epic:

> The story of Milton [*Paradise Lost*] might be told in two pages—it is this which distinguishes an Epic Poem from a Romance in metre. Observe the march of Milton—his severe application, his laborious polish, his deep metaphysical

> researches, his prayers to God before he began his great poem, all that could lift and swell his intellect, became his daily food. I should not think of devoting less than 20 years to an Epic Poem. Ten to collect materials and warm my mind with universal science. I would be a tolerable Mathematician, I would thoroughly know Mechanics, Hydrostatics, Optics, and Astronomy, Botany, Metallurgy, Fossilism, Chemistry, Geology, Anatomy, Medicine—then the mind of man—then the minds of men—in all Travels, Voyages and Histories. So I would spend ten years—the next five to the composition of the poem—and the five last to the correction of it. So I would write haply not unhearing of that divine and rightly-whispering Voice, which speaks to mighty minds of predestinated Garlands, starry and unwithering.[5]

Future whispers of glory notwithstanding, such an ambitious project seems almost doomed to failure from the start. If Coleridge insists on including everything, and on having all knowledge before getting started, he must fall short.

Many commentators on Coleridge have noted this encompassing tendency, which Thomas McFarland defines as fundamental to Coleridge and to his age:

> To understand Coleridge's thought, both in its own structure and in its relationship to the thought of his contemporaries, it is necessary to refer all its manifestations constantly and explicitly to the systematic unity, the total organicism which he, and almost all other thinkers of his era, accepted as the necessary condition of any intellectual activity at all.[6]

McFarland explains this tendency as a responsibility to vast amounts of data, based on a philosophical belief in complexity rather than simplicity, in accumulation rather than improvisation:

> If one tends to travel light intellectually, to live, as it were, out of a suitcase—after the manner of Wittgenstein, or Schlick, or even of Socrates—then no great housekeeping abilities are called for; but if one tends to admit intellectual responsibility for an enormous amount of data, with a continuing urge to accumulate still more, then the internal

> economy of this intellectual establishment becomes increasingly important. It is this principle of internal economy that we call system.[7]

As a philosophical system, such inclusiveness is more than admirable. It seems the apex of liberal, educated open-mindedness. What McFarland sees as Coleridge's central idea sets up a principle of inclusion that requires the thinker to include almost all possible systems of thought:

> The deeper ... we penetrate into the ground of things, the more truth we discover in the doctrines of the greater number of the philosophical sects.... all these we shall find united in one perspective central point, which shows regularity and a coincidence of all the parts in the very object, which from every other point of view must appear confused and distorted. The spirit of sectarianism has been hitherto our fault, and the cause of our failures. We have imprisoned our own conceptions in the lines, which we have drawn, in order to exclude the conceptions of others. [Coleridge then quotes Leibniz] J'ai trouvé que la plupart des sectes ont raison dans une bonne partie de ce qu'elles avancent, mais non pas tant en ce qu'elles nient. [my translation: I have found that most sects are quite correct in most of what they proclaim but not so correct in what they deny.][8]

The lack of inclusiveness arises from the exclusion of opposites from most systems of thought. Any partial truth is limited if it does not acknowledge the truths that oppose it, because it relegates those opposing truths to the category of falsehood:

> ... the most influencive Errors have ever been ... partial Truths mistaken for the whole Truth, Truths divorced from their correspondent and supporting opposites, and converted into contrary Falsehoods by being reciprocally unbalanced and disintegrated.... he alone deserves the name of a Philosopher, who has attained to see and learnt to supply the difference between Contraries that preclude, and Opposites that reciprocally suppose and require, each the other.[9]

Over and over again, Coleridge explicitly sets such open-minded inclusiveness as his goal:

> 'My system,' he told his nephew, 'if I may venture to give it so fine a name, is the only attempt I know ever made to reduce all knowledges into harmony. It opposes no other system, but shows what was true in each; and how that which was true in the particular, in each of them became error, *because* it was only half the truth.'[10]

Keats was one of the first to recognize the probable result of such an attempt at vast inclusiveness. In his definition of Negative Capability, he finds in Shakespeare an example to be admired, but in Coleridge an example to be lamented:

> several things dovetailed in my mind, & at once it struck me, what quality went to form a Man of Achievement especially in Literature & which Shakepeare posessed so enormously—I mean Negative Capability, that is when man is capable of being in uncertainties, Mysteries, doubts, without any irritable reaching after fact & reason—Coleridge, for instance, would let go by a fine isolated verisimilitude caught from the Penetralium of mystery, from being incapable of remaining content with half knowledge.[11]

While we can lament the negative result of Coleridge's system building, we can also praise what he did accomplish:

> The rich multi-level quality of Coleridge's imagination was obviously achieved at tremendous cost. It contains terrible tensions and contradictions.... the essential terms of Coleridge's reconciling system are dialectical. They stem initially from his awareness of contradictions within his own experience, ... between radical disbelief and traditional faith.... when he read Kant's *Critique of Pure Reason* ... Coleridge found the fundamental encroachment of the *subject* upon the *object* in human experience. In the *Aids to Reflection* he urged, as the greatest assistance to clear thinking, the re-introduction into English of 'subjective' and 'objective' reality—terms which are now in completely current use.[12]

Thus, within the concern for systematic wholeness, Coleridge insists on the inclusion of opposites. One of the primary sets of opposites of course is that between subject and object, which also can be defined as inner and outer:

> The notebooks record the collisions of a hugely developed sense of inner reality with a hugely developed sense of outer reality, with neither sense giving ground.[13]

As so often, Coleridge mocks his extreme interest in this issue, making it into a weakness that bores his listeners. He makes the issue extremely personal as he implicitly, like Keats, participates in all of existence. His delight in revealing differences (we shall later encounter the term *desynonomize*) is overcome by his delight in making connections, as the circling ripples of his imagination try to include all:

> I feel too intensely the omnipresence of all in each, platonically speaking, or psychologically my brain-fibres or the spiritual light which abides in the brain marrow, as visible light appears to do in sundry rotten mackerel and other smashy matters is of too general an affinity with all things. And though it perceives the difference of things, yet is eternally pursuing the likeness, or rather that which is common. Bring me two things that seem the very same, and then I am quick enough to shew the difference, even to hair-splitting; but to go on from circle to circle till I break against the shore of my hearer's patience or have my Concentricals dashed to nothing by a Snore, this is my ordinary mishap.[14]

Through these pathways of logic Coleridge continues to explore the question. Although he tries to penetrate the coincidence of opposites, the problem always remains a stand-off for him. As the foundation of his always promised opus maximum, it may have prevented him from completing that amibitious project:

> It was with logic as the focal point that Coleridge early began his investigation of the 'Coincidentia oppositorum,' the idea of the reconciliation of opposites. By 1803 he had formulated a detailed prospectus of his 'great work.'[15]

His project to fill a notebook with examples of "extremes meet" was based on his belief that all philosophy was contained in that phrase:

> Extremes meet—a proverb, by the bye, to collect and explain all the instances and exemplifications of which would constitute and exhaust all philosophy.[16]

This obsession with opposites dominates much of Coleridge's writing, both poetic and philosophical. In the *Biographia* he even defines contraries as the basis not only of philosophy but of all creation:

> the transcendental philosopher says; grant me a nature having two contrary forces, the one of which tends to expand infinitely, while the other strives to apprehend or *find* itself in this infinity, and I will cause the world of intelligences with the whole system of their representations to rise up before you.[17]

Thus, in Coleridge, as in Cusanus, the entire philosophical structure begins with a statement of contrary forces, centrifugal and centripetal.

Although Coleridge wants all, although nothing less than the whole Truth will satisfy him, again and again he finds reasons to back away from the holistic vision that attracts him. One of the most excruciating tensions of contraries that Coleridge felt himself caught in was the tension between pantheism and orthodoxy. His conversation poem "The Aeolian Harp," composed in 1795, illustrates that tension. After setting a scene of domestic bliss with cottage and wife, inspired like almost every Romantic poet by the Aeolian Harp, Coleridge suddenly imagines

> ... the one Life within us and abroad,
> Which meets all motion and becomes its soul,
> A light in sound, a sound-like power in light,
> Rhythm in all thought, and joyance every where—
> Methinks, it should have been impossible
> Not to love all things in a world so fill'd
> ...
> And what if all of animated nature
> Be but organic Harps diversely fram'd,
> That tremble into thought, as o'er them sweeps
> Plastic and vast, one intellectual breeze,
> At once the Soul of each, and God of all?
> But thy more serious eye a mild reproof
> Darts, O beloved Woman! nor such thoughts
> Dim and unhallow'd dost thou not reject,
> And biddest me walk humbly with my God.
> Meek Daughter in the family of Christ!

Well hast thou said and holily disprais'd
These shapings fo the unregenerate mind;
Bubbles that glitter as they rise and break
On vain Philosophy's aye-babbling spring.

As we see in this poem, Coleridge will have good reason to sympathize with the Berengarius (see discussion below) who had trouble perceiving the Real Presence in the Eucharist and who found himself struggling between conscience and the Pope. No matter how hard he tries in his poetry to achieve this oneness with all of life, Coleridge always falls back into isolation:

> The quest for a 'something one & indivisible' underlying and animating the world is perhaps above all others the unifying principle of Coleridge's multifarious writings, although it will be seen how his statement of the 'One Life' is persistently checked and qualified.[18]

Because of his sensitivity to his audience, in this case his wife Sara as spokesman for orthodox Christianity, Coleridge recants his pantheism almost as soon as it is spoken. For, like Bernegarius, whom I will discuss at the end of this chapter, he will not be able to live in his society, and certainly not as a clergyman, if he does not hold to the orthodox line. Ironically, seeing more unity than is officially allowed will place him into a form of excommunication, which will separate him from those closest to him instead of increasing the social oneness that is implied by a belief in pantheism.

Instead of blaming Coleridge for his inability to resolve this problem, McFarland defines the problem as existing in the nature of things rather than in Coleridge himself:

> Inability either really to accept or wholeheartedly to reject pantheism is the central truth of Coleridge's philosophical activity.... As with the dilemma of Hamlet, who, not indecisive *in himself*, is confronted with alternatives that *in themselves* admit of no right solution, so with the dilemma of Coleridge: he could not resolve the ambivalences of the *Pantheismusstreit* without diminishing one whole side of his awareness and vital commitment. And so he bore the pain of conflicting interests rather than choose

> the anodyne of a solution that did violence to the claims of either side in the conflict.[19]

Thus McFarland agrees with Holmes in attributing courage to Coleridge's failure to reconcile opposites. Indeed, in their formulations, Coleridge's refusal to let either side of the opposition win constitutes his admirable strength. McFarland thus paints Coleridge as braver than most, able to bear almost unbearable tensions because of his principled refusal to compromise.

Indeed, for all these writers who are concerned with the coincidence of opposites, the question of whether the problem exists in the very nature of things is an important one. Cusanus, Pope, and Blake definitely believe that the coincidence of opposites is fundamental to the make-up of the universe and of humanity. They assert that belief in various ways and stand by it. Coleridge, Mary Shelley, and Percy Shelley, to varying degrees, want to believe in the coincidence of opposites, but fail to maintain and assert consistently and strongly their vision in that direction. McFarland defines the split in Coleridge as one between head and heart:

> We are here interested in the *emotional* attraction of pantheism for Coleridge; on the *rational* level his attitude to pantheism is clear and unfailingly censorious.[20]

In "The Aeolian Harp" Coleridge reaches for pantheism in an attempt to obtain and reconcile all. Such a reconciliation would necessitate the joining together of philosophical opposites. Then he finds himself also compelled, with just as much force and in the reverse direction (like his very description of the forces of the universe quoted above) to reject that wholeness. Therefore he finds himself caught in a higher coincidence of opposites, one that vacillates between accepting and rejecting the coincidence of opposites.

In his late, politically conservative work, *On the Constitution of Church and State,* Coleridge combines his philosophical and religious ideas with political ones to propose what he calls the clerisy, a kind of mediating force made up of the intellectual estates of universities and schools in addition to the clergy, that would provide for the constructive balance of opposites in society:

> Coleridge presented this national clerisy as the great reconciling and sustaining body within the Constitution as a whole, which would balance those forces of permanency and progression which are continuously in conflict within the nation.... The clerisy would be the dynamic centre of renewal within national life, its object 'to secure and improve that civilisation, without which the nation could be neither permanent nor progressive.'[21]

In that work, Coleridge makes a distinction to clarify his conception of opposites. As with all these thinkers into opposites, some terms take on great importance, while others seem unimportant.

> Permit me to draw your attention to the essential difference between *opposite* and *contrary*. Opposite powers are always of the same kind, and tend to union, either by equipoise or by a common product. Thus the + and – poles of the magnet, thus positive and negative electricity are opposites. Sweet and sour are opposites; sweet and bitter are contraries. The feminine character is *opposed* to the masculine; but the effeminate is its *contrary*. Even so in the present instance, the interest of permanence is opposed to that of progressiveness; but so far from being contrary interests, they, like the magnetic forces suppose and require each other. Even the most mobile of creatures, the serpent, makes a *rest* of its own body, and drawing up its voluminous train from behind on this fulcrum, propels itself onward.[22]

Owen Barfield dismisses this distinction:

> The distinction between 'opposite' and 'contrary' made in ... *Church and State* ... may, I think, be ignored. In common use both terms are taken to connote mutual exclusion. Coleridge was there apparently attempting to 'desynonymise' them by appropriating this connotation to one of them ('contrary') only. The distinction however is not one that he maintained. While, in the footnote, 'contrary' is made virtually equivalent to 'contradictory,' elsewhere it is not infrequently synonymous with 'opposite.'[23]

Similarly, Blake asserts fundamental distinctions between negations and contraries, which Leopold Damrosch dismisses as inconsequential. In Blake, "contrary" is the favorable term: everything needs its contrary. In Coleridge "opposite" is the favorable term: opposites tend to union. In Blake, "negation" is the pejorative term. Negations try to cancel out the forces they feel opposed to. In Coleridge "contrary" is the pejorative term: contraries try to cancel each other out. Blake's negations are defined in terms of religious systems that try to impose their reifications onto others. Coleridge, however, simply finds logical impossibility in certain statements of opposites:

> Opposites, he well observes, are of two kinds, either logical, i.e. such as are absolutely incompatible; or real without being contradictory. The former he denominates Nihil negativum irrepresentabile, [Engell's footnote: "Nothing in a negative sense, not representable" (the logical opposite)—i.e. the state of a body both at rest and in motion, as C[oleridge] goes on to explain, following Kant] the connexion of which produces nonsense. A body in motion is something—Aliquid cogitabile; but a body, at one and the same time in motion and not in motion, is nothing, or at most, air articulated into nonsense. But a motory force of a body in one direction, and an equal force of the same body in an opposite direction is not incompatible, and the result, namely rest, is real and representable.[24]

This distinction between opposites and contraries, between dynamism and stasis, seems to haunt Coleridge's poetry. In his Dejection Ode, Coleridge is at least in part answering his friend William Wordsworth's Intimations Ode, which wonders why childhood bliss disappears:

> There was a time when meadow, grove, and stream,
> The earth, and every common sight,
> To me did seem
> Apparelled in celestial light,
> The glory and the freshness of a dream.
> …
> Whither is it fled, the visionary gleam,
> Where is it now, the glory and the dream?

Although Coleridge's memory is not of such perfect bliss, he does remember something similar to Wordsworth's memory:

> There was a time when …
> … hope grew round me, like the twining vine,
> And fruits, and foliage, not my own, seemed mine.

Coleridge does not specifically recall the glow in Nature that Wordsworth recalls; he had only hope. Whereas Wordsworth directly perceived something outside himself that gave him a feeling of warmth and joy, Coleridge's perception of joy outside himself was only a hope; even less than a hope, it only "seemed" to grow around, comfort, and belong to him. Finally, Coleridge receives no reciprocity; he has to perform the whole task himself:

> I may not hope from outward forms to win
> The passion and the life, whose fountains are within.
>
> O Lady! we receive but what we give,
> And in our life alone does Nature live

Coleridge sets himself the impossible task of virtually creating Nature by his own power. In that sense "All Nature is but Art" is an impossibly heavy burden.

His metaphor of marriage with Nature makes the herculean task even more painful:

> Joy, Lady! is the spirit and the power,
> Which, wedding Nature to us, gives in dower
> A new Earth and new Heaven.

Although the syntax becomes rather confusing, Coleridge here makes Joy the father of the bride. Only Joy can give the dowry of a new Earth and new Heaven. But where does that Joy come from? It can come only from within the poet, who in this metaphor is the bridegroom. Thus Coleridge sets up a short-circuit. Asking for a new Earth and a new Heaven from outside, he can imagine it only as coming from within himself, just as earlier in the poem, when trying to receive joy from Nature, he asserted that he could receive only what he gave. He has put himself in the impossible double position of groom and father of the bride! Therefore Coleridge is left with an insoluble dilemma, which degenerates further into "Reality's dark dream."

By contrast, when Wordsworth tries to answer the problem of joy in the last half of the Intimations Ode, he constructs two outside, benevolent forces with which to interact, one in Nature and one in the idealism of Platonic metempsychosis. He is grateful for obstinate questionings that give him an origin and a goal beyond Nature. While the homely Nurse tries to keep him as an inmate, Wordsworth's babe escapes because of the Platonic glory from which he came. Even though Wordsworth in remarks outside the Intimations Ode claimed that reincarnation was only a hypothesis, not his firm belief, the tone of the poem is strong and certain, not at all like Coleridge's doubts and hesitations. Wordsworth is able to thrust himself, by the willing suspension of disbelief, into a position that might not accord with his Christian orthodoxy, a move which, as we saw in "The Aeolian Harp," Coleridge is not able to sustain.

Thus Coleridge's poem implies a kind of stasis in the total system rather than a dynamic reciprocity. In both "The Aeolian Harp" and the Dejection Ode, Coleridge does not measure up to Wordsworth's dynamic recreation of the dead-ends of perception. Instead of setting up a reciprocal system, Coleridge thinks that glory can arise only from within. By putting the whole burden on himself, Coleridge remains in stasis, unable to move because he doesn't really seem to believe in a corresponding outside opposite. He has put himself in a position perilously close to that of the *Alastor* poet (see Chapter 5 on Percy Shelley), who ignores the outside world in preference to his solipsistic musings. He can not therefore imagine the opposites of self and other coming together, as do Blake and Shelley.

His philosophical theory finally does not fully enter his poetry. Wordsworth's tentative hypothesis strongly invests the Intimations Ode (and "Tintern Abbey") but Coleridge's strongly held belief remains tentative and hesitant in "The Aeolian Harp" and the Dejection Ode.

Like Cusanus, Coleridge begins his whole system with opposite forces:

> Now the transcendental philosophy demands; first, that two forces should be conceived which counteract each other by their essential nature; not only not in consequence of the accidental direction of each, but as prior to all direction, nay, as the primary forces from which the conditions of all possible directions are derivative and deducible: secondly, that these forces should be assumed to be both alike infinite, both alike indestructible. The problem will then be to discover the result or product of two such forces, as distin-

> guished from the result of those forces which are finite, and derive their difference solely from the circumstance of their direction. When we have formed a scheme or outline of these two different kinds of force, and of their different results by the process of discursive reasoning, it will then remain for us to elevate the Thesis from notional to actual, by contemplating intuitively this one power with its two inherent indestructible yet counter-acting forces, and the results or generations to which their interpenetration gives existence, in the living principle and in the process of our own self-consciousness.[25]

But it is this movement of the coincidence of opposites "from the notional to the actual" that Coleridge does not achieve in his poetry. Like Cusanus and like Pope, his reasoning from what he knows seems to hold him back. Blake's method of rejecting Reason as a method and going straight to intuition penetrates that limit. As long as we reason only from what we know, we are limited because, as Blake emphasizes in "There Is No Natural Religion," we start with certain pre-conceived definitions of what is available to our perception. Perceiving more than empiricism allows and thus refusing to allow the limits of empiricism and reason to control us, we can perceive miracles through expanded perceptions. Empiricism, like any deadening, abstract system, tells that we do not have the experiences that we have because they do not fit the official system.

Coleridge's famous definition of the Imagination does try to bring together the divine and the human, even while it distinguishes between them:

> The IMAGINATION then I consider either as primary, or secondary. The primary IMAGINATION I hold to be the living Power and prime Agent of all human Perception, and as a repetition in the finite mind of the eternal act of creation in the infinite I AM. The secondary I consider as an echo of the former, co-existing with the conscious will, yet still as identical with the primary in the *kind* of its agency, and differing only in *degree*, and in the *mode* of its operation. It dissolves, diffuses, dissipates, in order to re-create; or where this process is rendered impossible, yet still at all events it struggles to idealize and to unify. It is

> essentially *vital*, even as all objects (as objects) are essentially fixed and dead.
>
> FANCY, on the contrary, has no other counters to play with, but fixities and definites. The Fancy is indeed no other than a mode of Memory emancipated from the order of time and space; and blended with, and modified by that empirical phaenomenon of the will, which we express by the word CHOICE. But equally with the ordinary memory it must receive all its materials ready made from the law of association.[26]

First, Coleridge defines the human imagination as a repetition of the divine mind, not as an identical force, as will Blake. Second, although he claims that the secondary imagination is "identical with the primary," he clearly gives it a lesser place not only by the denomination "secondary," but also by saying that it differs in degree from the primary. It is clearly lesser on some kind of measurable scale. But the greatest diminishment in the definition comes when, after defining the secondary imagination as having the power to re-create, he immediately qualifies the marvelous power with a resounding, discouraging, "where this process is rendered impossible." So, after being relegated to a lower level in the scale of imaginations (but soon to be reassured by hearing that the fancy is even lower), the secondary imagination must find frequent frustration, of the kind that we have seen in "The Aeolian Harp." Struggling to idealize and to unify in that poem, Coleridge found himself forced to recant. And now in his definition of the marvelous power of the creative imagination, he finds himself forced to qualify to the point of frustration.

In her study of the imagination, Mary Warnock defines Coleridge's concept in very strong and constructive terms:

> something working actively from within to enable us to perceive the general in the particular, to make us treat the particular, whether something we see or something we call up as an image, as symbolic, as meaning something beyond itself.... Imagination ... must try to create one thing (one thought or one form) out of the many different elements of experience; and this entails extracting the essence of the differing phenomena of experience.[27]

This explanation, like Coleridge's definitions, implies the abstraction of total system, not a Blakean imagination of minute particulars. Most people naturally think that oneness, total unity can come only from abstracting, from generalizing about the characteristics that different particulars have in common, but the idea of the coincidence of opposites helps us break through that limitation to see unity and disparity simultaneously, just like seeing two parts of an optical illusion at the same time. Warnock reproduces an optical illusion that can be seen as *either* a rabbit *or* a duck.[28] As in the more famous optical illusion of two faces or a vase (reproduced in Stephen Prickett's *Words and the Word*),[29] most people think that they can see only one or the other, that their powers of generalizing and abstracting from vague lines must choose. (For more discussion of this optical illusion, see Chapter 7: Internal Eternity.) They also naturally think that words must abstract and generalize.

As Coleridge continues to explain his theory of imagination in the following chapter of *Biographia Literaria,* he continues to define it in terms of reconciling oppositions:

> The poet … diffuses a tone and spirit of unity … by that synthetic and magical power to which we have exclusively appropriated the name of imagination. This power, first put into action by the will and understanding, and retained under their unremissive though gentle and unnoticed control, reveals itself in the balance or reconciliation of opposite or discordant qualities: of sameness with difference; of the general with the concrete; the idea with the image; the individual with the representative.[30]

Deirdre Coleman and Peter Otto, in their introduction to *Imagining Romanticism,* deconstruct Coleridge's definition, by revealing how fancy apparently undercuts imagination:

> What is not so often remarked is that the *Biographia* passage, in the course of mapping the progression of the 'I AM' to the primary and then to the secondary imagination … maps a strange countermovement to its own argument. Coleridge's affirmation of presence is shadowed by a regressive movement of inversion and loss.... Repetition and reduction suggest that the primary imagination is engendered by a process which more closely parallels the repeti-

> tions of fancy than the vital and idealizing activities that it is itself said to undertake: the vital is born from the mechanical; the original finds its moment of birth in the secondary and derivative.[31]

But this apparent undercutting of Coleridge's idea, if seen from the perspective of the coincidence of opposites, actually strengthens it. For, as Blake will make even clearer, the dynamism of the coincidence of opposites always produces reversals of this kind. When making distinctions, or desynonymizing, Coleridge is not separating words into discrete categories that never interact. Rather, he is setting up a dynamic movement back and forth between opposites.

When Stephen Prickett analyzes Coleridge's attempts to desynonymize words, he also deconstructs Coleridge's approach and then seems surprised when he discovers that Coleridge is trying to re-synonomize with another coined term, "esemplastic." We can learn from the coincidence of opposites not to be surprised at this kind of maneuver, and recognize it from the beginning. Or perhaps the power of the coincidence of opposites is always to surprise, always to keep the dynamics of our awareness of world and words alive.

When Prickett traces Coleridge's defition of imagination back to Greek thought and to the word *logos*, he concludes:

> For Heraclitus, for instance, the logos, or God, was seen as the common connecting element in all extremes.... For [Coleridge] the biblical logos is part of an even earlier philosophic tradition where the 'Word' combines under tension opposite or discordant qualities in a creative unity.[32]

Prickett is uncovering the fundamentally religious foundation of Coleridge's coincidence of opposites:

> there is almost a suggestion that the Primary and Secondary Imaginations are, as it were, typologically linked in a manner analogous to the Old and New Testaments: that is, that we only fully come to understand the unconscious activity of sense-perception through the conscious activity of artistic creation.[33]

This insight by Prickett is very similar to one that I will develop later in Blake. For now, we can recognize that in Prickett's reading the primary imagination, like the Old Testament God, begins all life, but does not fully realize its potential. The secondary imagination, like the New Testament God, fulfills His potential by sending His son to link the realms of divine and human, in a move similar to the one insisted on by Cusanus. Without Jesus as the divine humanity, according to Cusanus and Coleridge and Blake, the gap between divine and human remains unbridgeable.

The source of the whole of Coleridge's definition is the primary imagination of God, the great "I AM," existence itself. Without that fiat, there would be no further commands such as "Let there be light." Without the great "I AM," who is able to contain all opposites, there would be no existence. Although from God's point of view, as it were, the imagination is the repetition of the great I AM, from the human perspective, the imagination is based on two forces, self (centripetal) and other (centrifugal):

> C[oleridge] summarizes an outline or schema of a definition of the imagination that has clear similarities to long deductions and definitions of the imagination in Fichte and Schelling. Two forces or concepts in dynamic tension both find themselves in the imagination, which reconciles and unifies them: the self or mind ("I am") with nature or the cosmos, the subjective with the objective.[34]

Logic and social injunction cannot dissolve the boundaries between man and nature, or between man and God; in fact they strengthen those boundaries. Only the imagination, which dissolves and dissipates in order to re-create, can overcome the boundaries. In Pope there was no real attempt to overcome the boundaries. The distinctions were crucial, fundamental to humans, and only from the mysterious power of God's viewpoint did the opposites coincide. In Blake and Shelley and Coleridge, the human participates in the conjunction of opposites not just by accepting the mystery imposed from on high, but by participating in the divine as fully as in the human. Coleridge often seems to wish he could attain that height; Shelley and Blake actually accomplish the feat.

One of the strangest characteristics of Coleridge's definition of the imagination is its placement in the *Biographia*. It comes right at the end of Chapter XIII, right at the end of Part I, with a promise:

> I shall content myself for the present with stating the main result of the Chapter, which I have reserved for that future publication, a detailed prospectus of which the reader will find at the close of the second volume.[35]

As Engell drily remarks, "The prospectus does not appear in the 1817 ed and apparently was never written."[36] And what made Coleridge decide to state only the main result—the definitions of imagination and fancy—and save the rest for a future that never came? It was "a letter from a friend, ... [a] very judicious letter, which produced complete conviction on my mind,"[37] a conviction that appears in the quotation above.

Just before the "letter from a friend," Coleridge has been building his definition based on the interpenetration of opposite forces:

> The counteraction then of the two assumed forces does not depend on their meeting from opposite directions; the power which acts in them is indestructible; ... and as something must be the result of these two forces, both alike infinite, and both alike indestructible; and as rest or neutralization cannot be this result; no other conception is possible, but that the product must be a tertium aliquid, or finite generation. Consequently this conception is necessary. Now this tertium aliquid can be no other than an inter-penetration of the counteracting powers, partaking of both. [Coleridge's page contains asterisks][38]

At this point a line of asterisks interrupts the headlong momentum, and the Coleridgean short-circuit kicks in. Just as in "The Aeolian Harp," when orthodoxy denies pantheism; just as in the writing of "Kubla Khan," when the person from Porlock interrupts the poet's recall of his reverie; just as we shall see later in his understanding of Berengarius; Coleridge always blames outside forces for his inability to see the real presence, the mystical wholeness. He puts himself into an impossible bind: on the one hand, as in the Dejection Ode, he must re-create all of the outside world single-handedly; on the other hand, he can not finish his writing because of the interruptions and misunderstandings of the outside world:

> Thus far had the work been transcribed for the press, when I received the following letter from a friend ... [ellipsis in original] Dear C. You ask my opinion concerning your Chapter on the Imagination, both as to the impressions it made on myself, and as to those which I think it will make on the PUBLIC, ... As to myself, and stating in the first place the effect on my understanding, your opinions and method of argument were not only so *new* to me, but so directly the reverse of all I had ever been accustomed to consider as truth, that even if I had comprehended your premises sufficiently to have admitted them, and had seen the necessity of your conclusions, I should still have been in that state of mind, which ... you have so ingeniously evolved, as the antithesis to that in which a man is, when he makes a *bull*. In your own words, I should have felt as if I had been standing on my head.[39]

This interruption, which was composed by Coleridge himself (and don't some of us suspect as much about the person from Porlock?), stops the idealistic rush just as surely as the reproving glance of Sara quells the unregenerate musings in "The Aeolian Harp." But instead of being thwarted by something from outside, as Coleridge makes it appear, he is being thwarted by something inside. In a bizarre revisiting of the Dejection Ode from a different angle, Coleridge tries to make us believe that he can not project his joyful theory because the cruel world will throw it right back at him as if in a rejecting mirror. In that poem, he could not receive any joy from nature because he could receive only what he gave. In this passage from the *Biographia* he can not receive the joy of a delighted audience because he will not give it.

The spurious letter states as its primary concern the impression that Coleridge's definitions will have on the reading public. While this concern for audience is certainly a legitimate one, Coleridge is going to great pains to fabricate a resistant reader to interrupt his attempt to explain his theory to the general public, whose reaction he fears. His fictitious interlocutor objects to the theory, among other reasons, because it is "the reverse" of what he has previously believed. Thus Coleridge fears that his readers will use their ordinary, everyday logic and refuse to admit anything that contradicts it, especially a theory that wants them to accept a coincidence of opposites. His theory, he thinks, is so opposed to the prevailing ideology that it has no chance of acceptance. The cousin of the person from Porlock then goes on to explain that instead of making a bull,[40] he feels that he has received one that has stood him on his head. In other words, Coleridge's pronouncements, like Cusanus's, are usually seen as Lewis Carroll-like jokes that disturb good, decent folk instead of revealing fundamental truths.

When Owen Barfield elucidates Coleridge's thought, he emphasizes the concept of "polarity," which was so important that Coleridge invented a short-hand symbol for it:

> he habitually employed the symbol)-(, to avoid the tedium of writing out in full some such phrase as 'is polarically related to' or 'is the polar opposite of.'[41]

According to Barfield's explanation of Coleridge's ideas, the importance of polarity is not just its usefulness or its applicability or its potential for the imagination to use; it is a law of the universe, a fundamental fact of existence:

> Polarity is, according to Coleridge, a 'law'; it is a law which reigns through all Nature; the duality of the 'opposite forces' is the *manifestation* of a prior unity; and that unity is a 'power.' It is not, that is to say, any abstract 'principle of unity' or of identity—a point which it is hardly possible to over-emphasise, since that is precisely what it is commonly presumed to be....
>
> Polarity is dynamic, not abstract. It is not 'a mere balance or compromise,' but 'a living and generative interpenetration.' Where logical opposites are contradictory, polar opposites are generative of each other—and together generative of new product. Polar opposites exist by virtue of each other *as well as* at the expense of each other; 'each is that which is called, relatively, by predominance of the one character or quality, not by the absolute exclusion of the other.' Moreover each quality or character is present *in* the other. We can and must distinguish, but there is no possibility of *dividing* them.

> But when one has said all this, how much has one succeeded in conveying? How much use are definitions of the undefinable? The point is, has the imagination grasped it? For nothing else can do so. At this point the reader must be called on, not to think about imagination, but to use it. Indeed we shall see that the apprehension of polarity is itself *the basic act of imagination.*[42]

Barfield insists on activity and not just contemplation:

> what does the principle of contradiction, without more, offer to a mind? What does the mind obtain by 'submitting all positions alike ... to the criterion of the mere understanding'? Quite literally nothing. The principle of contradiction tells us nothing of what nature, or anything else, *is*. It tells us only what it is *not*; and, in doing so, clenches our absurd detachment from it.[43]

Although Barfield is partly correct here, he does give a misleading impression. To the extent that the coincidence of opposites means submitting all positions to a criterion of understanding, or reason, it is only negative. Often opposites in Coleridge do simply negate each other. In "The Aeolian Harp" the poet's reverie is stopped dead by the reasoning contradiction of orthodoxy spoken by Sara. But in a more dynamic use of the coincidence of opposites, as in Blake, the contradiction which affronts mere reason encourages us to expand our imaginations and see more than limited reason allows. The obvious result of contradiction is the telling us what is not, but the larger result is telling us what is, in the largest possible sense. Monism can easily degenerate into exclusive dogmatism. But the coincidence of opposites allows us to rise above preconceptions to a larger view. While telling us what is not, and keeping us from reifying, the coincidence of opposites also allows us the leap of imagination which sees the universe whole, which can see both the vase and the faces in the optical illusion, which can see both the human and the divine, both the living and the dead, which can overcome the limitations of the reason to see into the divine world.

In a description of metaphor, Barfield captures precisely the power of the coincidence of opposites. It is not just an abstraction, but a basic feature of existence and of language: seeing the coincidence of opposites is recapturing a kind of primitive innocence, but is also participating in a wholeness which reason forbids. Trying to insist on no coincidence of opposites denies a fundamental fact of existence.

> The element of contradiction is most apparent in that particular form of symbolic utterance called a metaphor; but it is certainly (for the understanding) characteristic of figurative language as a whole. Primitive language is instinctively figurative. The further back we penetrate in the history of speech, the more symbolic, and therefore the less logical, it shows itself to have been. There are, for instance, Hebrew roots, whose semantic range is so wide that one end of their gamut of meanings appears to us to be positively the reverse of the other. Living opposites have not yet been reduced to contradictions.[44]

In 1827 Coleridge published a poem on the subject of the monk Berengarius, who was the focus of an eleventh-century controversy that helped the Church clarify its position on the Real Presence in the Eucharist. Berengarius denied the Real Presence. Like many in the ongoing argument before the Church pronounced the final word, Berengarius believed in a symbolic, spiritual presence of Christ in the bread and wine rather than a physical one. During his struggles with the Church over this issue, he signed three professions of faith (in 1055, 1078, and 1079), each one more explicitly acknowledging the Church's orthodox position. Coleridge's poem sympathetically captures the uncertainty, the difficulty of seeing a mystery that the mind confesses to, but which the understanding refuses to grasp.

According to J. Robert Barth, Coleridge himself remained firmly in line with Anglican doctrine on the Eucharist, while arguing against Catholic beliefs:

> In his occasional comments on the sacrament of the Eucharist, Coleridge remained faithful ... in his consistent assertion of the real presence of the Body and Blood of Christ in the sacrament. For the mature Coleridge, this was never in doubt. What did occasionally attract his attention was the matter of *how* Christ is present in the Eucharist.[45]

This issue of *how* the Real Presence entered the bread and wine was indeed the focus of much of the Berengarian heresy.

Coleridge himself had no strong objection in principle to the Catholic doctrine of transubstantiation, but saw no need for such a doctrine and no clear biblical justification. Like Berengarius, he could not accept that the physical body born of Mary was the same one that appeared in the Eucharist: "applied to the phenomenal flesh and blood, it is nonsense."[46] But clearly, like the Berengarius of his poem, Coleridge did not think the matter worth fighting over. For Anglicans, the mystery of the Real Presence remains a mystery; how it happens is not prescribed by doctrine. For Catholics, however, and particularly for Berengarius, failure to see the Real Presence in the prescribed way can result in excommunication and eternal damnation.

LINES

Suggested by the Last Words of Berengarius
Ob. Anno Dom. 1088

No more 'twixt conscience staggering and the Pope
Soon shall I now before my God appear,
By him to be acquitted, as I hope;
By him to be condemned, as I fear.—

Reflection on the Above

Lynx among moles! had I stood by thy bed,
Be of good cheer, meek soul! I would have said:
I see a hope spring from that humble fear.
All are not strong alike through storms to steer
Right onward. What? though dread of threatened death
And dungeon torture made thy hand and breath
Inconstant to the truth within thy heart!
That truth, from which, through fear, thou twice didst start,
Fear haply told thee, was a learned strife,
Or not so vital as to claim thy life:
And myriads had reached Heaven, who never knew
Where lay the difference 'twixt the false and true!

Ye, who secure 'mid trophies not your own,
Judge him who won them when he stood alone,
And proudly talk of recreant Berengare—
O first the age, and then the man compare!
That age how dark! congenial minds how rare!
No host of friends with kindred zeal did burn!

No throbbing hearts awaited his return!
Prostrate alike when prince and peasant fell,
He only disenchanted from the spell,
Like the weak worm that gems the starless night,
Moved in the scanty circlet of his light:
And was it strange if he withdrew the ray
That did but guide the night-birds to their prey?

The ascending day-star with a bolder eye
Hath lit each dew-drop on our trimmer lawn!
Yet not for this, if wise, shall we decry
The spots and struggles of the timid Dawn;
Lest so we tempt th'approaching Noon to scorn
The mists and painted vapours of our Morn.[47]

Allan John Macdonald explains Bernegarius's heresy in a way that is reminiscent of the disinction between Aristotle's Law of Non-Contradiction, based on reason and nature, and Cusanus's *coincidentia oppositorum*, based on a higher divine law. Macdonald explains what is

> the fundamental principle of the whole Berengarian criticism of the Catholic dogma of Transubstantiation. It is contrary to the evidence of the senses.... Secondly, reason compels an admission that it is contrary to the law of nature that one thing can be changed into another without the breaking-up or annihilation of its original elements.[48]

So the issue joined by Berengarius pitted a rational explanation against a mystical one. The Church's position was that the opposites of bread/wine and flesh/blood do coincide in the Eucharist. Just as the mystery of the Incarnation made a supposedly impossible combination of spirit and flesh in the person of Jesus, so does the mystery of the Eucharist bring together the food and the risen body. Although we can still see the bread and wine, they are in reality the body of Jesus. But since we can still see the bread and wine, reasoned Berengarius, the doctrine of the Real Presence requires us to deny the evidence of our senses.

As Coleridge explores the issue, he sets up a carefully constructed opposition: to believe or not to believe. Whatever the intricacies of the theological arguments, Berengarius is torn between obedience to the order of the Pope to conform and the decision of his own conscience. The two sides of the opposition can not possibly be reconciled on this earth, but they *may* be reconciled in heaven. Berengarius has been condemned by the Pope for not being able to see the Real Presence in the Eucharist. Opposed to this condemnation, Berengarius's own conscience would acquit him. Berengarius hopes that God will agree with Berengarius's conscience and acquit him also, but his fear of God's condemnation, following the Pope's, is set in contrast to his hope of acquittal by his conscience. For indeed, his unstated fear is that even his own conscience may not acquit him; he might die and find that he has to condemn himself. His hope of acquittal is based on a perspective above the earthly; perhaps, for some reason that he can not see on earth, God, like his conscience, will acquit him. Here and now he is painfully balanced, "staggering" between conscience and Pope.

His deepest fear is that the two sides are irreconcilable. After death he will be in the presence of God, who may provide the missing element that will bring together the opposing earthly forces. Perhaps, Berengarius implicitly hopes in Coleridge's poem, God can leap over the dead-end that stymies him and the Pope. The hopeful formula coincides exactly with those of Cusanus discussed in Chapter 1: what appears to be in opposition on earth actually coincides in the mind of God. Cusanus's certainty is missing here, of course; the coincidence of opposites is merely a projected hope into the future. As in the Dejection Ode, the hope is not a Wordsworthian certainty. But Berengarius's only way out of the unbearable tension is to die and find that the opposites no longer will tear him apart. The poem implies that a higher resolution is possible, and hoped for, but the fear is based on the fact that the coincidence of opposites can not happen on earth, where Aristotle's Law of Non-Contradiction reigns.

Ironically, the hope that opposites can coincide in God in a way that they can not on earth is embodied in the very problem of Berengarius's heresy: his inability to acquiese to the doctrine of the Real Presence. He can not accept the church's doctrine that in the Eucharist the body and blood of Christ actually appear in the bread and wine. As we have seen in Coleridge, in "The Aeolian Harp" and in the Dejection Ode, he is caught in an impossible bind, impossible because of the constraints of logic.

Although this logic is not explicitly discussed in the lines to Berengarius, the dilemma certainly suggests that logical trap. If Berengarius is right, and divine and human do not become a perfect coincidence of opposites in the Real Presence of the Eucharist, then Berengarius has strong reason to doubt that God can solve the problem of the opposition of conscience and Pope by somehow bringing them together in a miraculous, divine coincidence of opposites. But if God can indeed effect a coincidence of opposites when Berengarius appears before him in heaven, then surely he can effect a coincidence of opposites in the body/blood and bread/wine of the Eucharist. If God can not make flesh and spirit coincide on earth, then he might not be able to resolve the problem of perceiving that paradox when Berengarius gets to heaven.

Berengarius is being judged by the Church because he cannot see; he cannot see the Real Presence that the Church demands that he see. The metaphor of sight guides the poem from the beginning: the lynx is legendary for its superior sense of sight, while moles are blind from living underground. Coleridge is thus performing another twist on the themes of opposition: although Berengarius's official heresy was that he could not see, Coleridge defines him as better at seeing than his contemporary accusers were.

In line 3, in Coleridge's apostrophe to Berengarius ("had I stood by thy bed" also means if I had been you, and in a fundamental way, I am like you) he attempts to take a reconciling view, that is to imagine what God's position on the matter might be. By seeing a hope springing from the fear, Coleridge makes fear yield its opposite.

As another part of his reassurance, Coleridge continues the separation that had been set up in Pope/condemnation/fear vs. conscience/acquittal/hope. He adds "hand and breath" to the former set of terms and "truth within thy heart" to the latter set. Now the latter set of terms receives most of the value, whereas before the two sets of terms had been equally balanced. The condemnation by the Pope is further devalued by being linked to "dungeon torture," whereas the acquittal of conscience is linked to "truth." So Coleridge has set up an equation in which heart equals spiritual equals acquittal while hand and breath equal physical equal condemnation.

Line 13 implies that the ignorant, who do not trouble their brains with the question of the Real Presence, do go to Heaven without worrying about it. In fact, the whole problem seems to be that Berengarius was too smart: if he had not ever worried about the Real Presence, as ordinary people do not worry, then he could have gone to heaven along with them. And, in line 9, if he had not possessed fear, he might have defied the Pope and gone to hell. His learnedness would have been better off as ignorance. He is better off if his potential courage becomes fear instead. Coleridge obviously sympathizes deeply with someone whose fears wisely overcame his bold reaches into heterodoxy, someone, who, like the Coleridge who took on responsibility for all knowledge, was just too smart for his own good.

In his intelligence, Berengarius had to go through the agony of twice recanting and finally dying without really knowing the answer to his problem before death. Berengarius, like Coleridge, from being too intelligent and too scrupulous, by considering too curiously, has worried about "the false and true," but has finally decided that the issue of the Real Presence is not important enough to take a stand on. The fear of earthly torture may have saved Berengarius from the eternity of hellish torture. The Inquisition wins again, just as in "The Aeolian Harp."

Whereas the first stanza of the poem addresses Berengarius himself, the second stanza addresses Bernegarius's modern-day accusers. By implication, they are parallel to the Pope who condemned Berengarius. Since they do not understand the perspective of Berengarius (and by implication do not understand that of Coleridge) those who judge are not seeing from God's perspective, which might very well include the coincidence of opposites. Most ironically, their accusation, like the Pope's, arises because they accuse Berengarius of not seeing from God's perspective when he does not see the Real Presence.

After the first stanza addresses Berengarius and the second stanza addresses his accusers, the third stanza speaks in first person plural, including all, especially Coleridge's contemporaries. In this third stanza Coleridge speaks from a lengthy historical perspective, as if he were trying to see from God's point of view. With the greatest tolerance and a vast sweep of time, Coleridge tries to reconcile the agonized oppositions of the beginning of the poem.

In line 29 the sun of enlightenment, brighter in the nineteenth century than in the eleventh, rises and removes darkness. The earlier starless night had been broken only by the single light of Berengarius. The sequence of dawn, morn, noon sets up a time-line of progress. The dawn was Berengarius's time, with only incipient light. By Coleridge's time, the morn, more light appears, but even though we can see better than did previous ages, we should not condemn Berengarius, who did his best with his limited light. And if we do judge Berengarius harshly, says Coleridge, then we can expect that the brightest time, the noon that will arrive in the future, will also judge us. Like Berengarius, we can hope to suspend earthly judgment and obtain heavenly acquittal. Since God will have forgiven, so should we, all the more so because Coleridge feels himself in the same problem when he tries to see the everlasting presence in Nature.

Coleridge, after imagining the prayer said by Berengarius before his death, declares his sympathy for one who would declare what he fain would see, even if he couldn't see it. As in "The Aeolian Harp," Berengarius/Coleridge sees more than others do, but his ideas do not fit the established system's doctrine. Over and over he attempts to see a coincidence of opposites that he cannot see, falls back from his position, confesses his inability, but then tries to remain orthodox nonetheless. As contrasted with the definite assertions of the coincidence of opposites in Shelley and Blake, Coleridge always hesitates. Therefore he understands and sympathizes with the apparent weakness of Berengarius.

Raoul Heurtevent, in his study of the heresy of Berengarius (without reference to Coleridge), introduces an argument very similar to Coleridge's: according to Heurtevent, in an age when an issue is not questioned, the language is necessarily less precise.[49] Thus, before Berengarius questioned the Real Presence, the Church had not worked to define the issue precisely. After the language has been made more precise because of controversy, it is not fair to implicate someone by means of that precise language without respect for its earlier, more naive context. Coleridge's similar point also calls for open-mindedness: in our more tolerant and enlightened time, we can better understand the dilemma of Berengarius, who saw more clearly than his contemporaries, although they thought that he could not see.

The issue between Berengarius and the Church was precisely the same as the issue between Coleridge and Nature: a distinction between what the senses require and what faith requires.

> Il [Florus, one of the debaters in the Berengarian issue] distinguait nettement les deux cotés de la question eucharistique. Ce que voient les sens, et ce que disent la raison et la foi ne concordent pas. Quant à l'explication, elle est simple: c'est un mystère.[50]
>
> [my translation:] He clearly distinguished the two sides of the question of the Eucharist. What the senses see and what reason and faith say do not coincide. Why? Simple. It is a mystery.

Finally, the matter can be referred back to the first principle that I discussed in Nicolas Cusanus, who refused to let the power of God be limited by Aristotle's Law of Non-Contradiction:

> all sections of Berengarian opinion are agreed that the accidence of bread and wine remains after consecration, ... By making this denial [that bread and wine can also be flesh and blood], the Berengarians leave out of account the will of God, for, according to the Psalmist, God can do anything that He wills, and if God is almighty He can will a change to take place in natural objects.[51]

Thus, despite his yearnings for wholeness, expressed especially in his philosophical and political prose, Coleridge in his poetry, specifically in "The Aeolian Harp," the Dejection Ode, and "Lines Suggested by the Last Words of Berengarius," chooses the limits of Aristotelian reason and Christian orthodoxy over the callings of his beloved *coincidentia oppositorum*. He had good reason to; most of us make the same kind of decison. His brilliance, the bright noon of his mind and soul, remind us of the scale of the struggle:

> Coleridge lives and gains more life with each generation, not because he completed metaphysical systems, like Schelling, Fichte, and Hegel, whose Procrustean achievements only feebly command our attention today; and not, as Dr. Leavis and T.S. Eliot and a hallowed tradition of Anglo-American commentary would have it, because he was a poet and critic in spite of his metaphysical preoccu-

> pations; but because he honoured to the full the demands of a reticulated response to all the data of his consciousness—because, in short, his was a mind of rare integrity. He did not, like Goethe, stifle his metaphysical interests the better to breathe in the green world of things, and he did not, like the transcendental systematists, distort the texture of experience to achieve a completed network of abstraction.
>
> Coleridge's endeavour was always towards system. But this orientation was first of all the need to connect rather than the need to complete.[52]

In the four years of his life remaining after the lines to Berengarius, Coleridge pursues a new train of thought that leads us on to Blake:

> During the last few years of his life, Coleridge seems to have reflected often on a Christological doctrine which, until then, had held no special place in his thinking: the doctrine of the Mystical Body of Christ.... In Notebook 41 ... "Christ [must] be considered not only as a spiritual divine Man but as the essential Divine Humanity."[53]

Had he lived to explore the issue more fully, Coleridge would have certainly, like Pope, chosen conservative, orthodox paths. The Berengarian tension of defying established authority he obviously chose to solve in a Berengarian fashion: recant all heresies, live in orthodoxy. Blake will explode those cautious fears, living to the full the implications of the Human Form Divine, maintaining both his orthodoxy and his defiance in a white-hot *coincidentia oppositorum*.

Notes

1. N.M. Goldsmith, "A Reconciliation of Opposites: Concepts of the Mind in Pope and Coleridge," *Prose Studies*, 7 (1984), p. 6.

2. Although we do not know whether Pope had read Nicholas of Cusa, we do know that Coleridge had. In his notebooks (ed. Kathleen Coburn, 1957–71) in 1803, we find the follow-

ing entry in the context of musings over "the nature of Being which Creatures possess":

> vide Cusan. Dialog. de Genesi, quomodo idem identificando pluralitatem producit.—<The Hebrews called God "Space."—> (I: 1379)

This cryptic note, although it does not tell us very much, does tell us that Coleridge thought that Cusanus was an important reference in the question of one of the fundamental aspects of the coincidence of opposites: the Many and the One.

3. *The Friend*, Second Section, Essay IV, 1818, in *The Portable Coleridge*, pp. 339–40.

4. Milton tried to outdo Homer; Wordsworth tried to outdo Milton:

> For I must tread on shadowy ground, must sink
> Deep—and, aloft ascending, breathe in words
> To which the heaven of heavens is but a veil.
> All strength—all terror, single or in bands,
> That ever was put forth in personal form—
> Jehovah—with his thunder, and the choir
> Of shouting Angels, and the empyreal thrones—
> I pass them unalarmed. (Prospectus to *The Recluse*)

5. *Collected Letters of Samuel Taylor Coleridge*, ed. Earl Leslie Griggs (Oxford, 1956), I:320.

6. Thomas McFarland, *Coleridge and the Pantheist Tradition* (Oxford: Clarendon, 1969), p. xxxix.

7. McFarland, p. 139.

8. *Biographia Literaria*; qtd. in McFarland, p. 135.

9. *Notebooks*; qtd. in Wheeler, p. 54.

10. *Table Talk*; qtd. in Richard Holmes, *Coleridge* (Oxford UP, 1982), pp. 51–52.

11. *Letters of John Keats*, ed. Robert Gittings (Oxford, 1970), p. 43.

12. Richard Holmes, *Coleridge* (Oxford UP, 1972), pp. 51,52.

13. McFarland, p. 111.

14. *Notebook*, Dec 1804, no. 2372; qtd. in Mary Warnock, *Imagination* (Berkeley: Univ. of California Press, 1976), p. 80.

15. Wheeler, p. 50.

16. *The Friend*; qtd. in Wheeler, p. 57.

17. *Biographia Literaria*, ed. James Engell and W. Jackson Bate (Princeton UP, 1983), 1:296.

18. Nigel Leask, *The Politics of Imagination in Coleridge's Critical Thought* (NY: St. Martin's, 1988), p. 23.

19. McFarland, p. 107.

20. McFarland, p. 120.

21. Holmes, *Coleridge*, p. 64.

22. *On the Constitution of Church & State*, ed. John Barrell (London: Dent, 1972), p. 16.

23. Owen Barfield, *What Coleridge Thought* (Middletown, CT: Wesleyan UP, 1971), p. 201.

24. *Biographia*, 1:298.

25. *Biographia*, 1:299.

26. *Biographia*, 1:304–05.

27. Warnock, pp. 83–84, 92.

28. Warnock, p. 185.

29. Stephen Prickett, *Words and the Word: Language, Poetics and Biblical Interpretation* (Cambridge UP, 1986), p. 163.

30. *Biographia*, 2:15–17.

31. Coleman, Deirdre, and Peter Otto, "Introduction," *Imagining Romanticism* (West Cornwall, CT: Locust Hill Press, 1992), p. x.

32. Prickett, *Words*, p. 15.

33. Prickett, *Words*, p. 17.

34. *Biographia*, 1:299.

35. *Biographia*, 1:304.

36. *Biographia*, 1:304.

37. *Biographia*, 1:300, 304.

38. *Biographia*, 1:300.

39. *Biographia*, 1:300.

40. *Oxford English Dictionary*: "1803 Sydney Smith: apparent congruity, and real incongruity of ideas, suddenly discovered."

41. Barfield, p. 34.

42. Barfield, p. 35.

43. Barfield, p. 110.

44. Barfield, p. 232.

45. Barth, *Doctrine*, p. 175.

46. Qtd. in Barth, *Doctrine*, p. 177.

47. Coleridge, *Poetical Works*, ed. E.H. Coleridge (1912; rpt. Oxford, 1978), pp. 460–61.

48. Macdonald, p. 256.

49. Raoul Heurtevent, *Durand de Troarn et les Origines de l'Heresie Berengarienne* (Paris: Beauchesne, 1912).

50. Heurtevent, p. 184.

51. MacDonald, p. 341.

52. McFarland, p. 110.

53. Barth, *Doctrine*, p. 136.

Chapter 4

Peering Through the Chinks: Selflessness and Narrative Structure in *Frankenstein*

Unlike Cusanus, Pope, and Coleridge, Mary Shelley does not explicitly seek to make opposites coincide in a larger whole, but she does implicitly seek a coincidence of opposites in *Frankenstein* when she constructs her narrative technique of concentric circles. On the simplest level, the largest circle of narration, that of a reader of the novel, contains all the smaller circles, which center on the creature's tale of the family whom he watches through a tiny chink. Our attention as readers becomes increasingly focussed on smaller and smaller circles until we fall into the danger of egotism as intense as that of Victor Frankenstein. But working contrapuntally against that increasing smallness caused by centripetal force, the narrative structure of *Frankenstein* creates an equal and opposite centrifugal force of the expanding circumference of selflessness, love, and disinterest.

At the structural center of the novel *Frankenstein,* the creature who has been created by Victor Frankenstein hunches in his hovel, which is attached to "a cottage of a neat and pleasant appearance."[1] In this hovel whose walls full of chinks allow the cold winds to penetrate, the creature finds one "small and almost imperceptible chink through which the eye could just penetrate" (92) into the adjoining cottage. Through this chink he observes a family from whom he learns speaking, reading, and other human behavior.

The creature's narrative, which tells his own story and that of the De Lacey family, is enclosed within the narrative which his creator, the scientist Victor Frankenstein, recites to a sea captain, Robert Walton. In turn Walton, who is sailing for the North Pole, tells his story of finding Frankenstein in letters which he writes to his sister, Mrs. Saville, back in England. Thus the story of the De Laceys is contained within the story told by the creature within the story told by Victor Frankenstein within the story written by Robert Walton within the story written by Mary Shelley. Furthermore, a preface, purportedly by the author of *Frankenstein* but actually by Percy Bysshe Shelley, instructs the reader in how to read the book. Mary Shelley's own introduction tells us how the story was conceived. So even if we do not count the cloth covers or the paper ones surrounding the book, even if we do not count the *Frankenstein* films to which the twentieth century has given birth, we still find no fewer than seven framing chinks to peer through.

In fact, as we actually read the novel, this structure is much easier to navigate than its description might suggest. We probably even forget, while we read the creature's tale, that it is enclosed within Frankenstein's and Walton's tales. The immediacy of tone in each narration invites us to become the audience of each narrator: first we are readers of the author's preface and introduction; then we become Walton's sister as she reads her brother's letters, then we become Walton as he listens to Frankenstein, then we become Frankenstein as he listens to the creature. At the same time we may find ourselves identifying with each narrator in turn. We may feel ourselves entering into the consciousnesses of Mary Shelley, Robert Walton, Victor Frankenstein, and the creature as they tell their stories. Thus as readers we may play several roles at once: while we are reading the creature's tale, we may identify with him. At the same time we may remember that the creature's tale is being told through Frankenstein's memory and that at the same time Frankenstein's tale is being told through Walton's memory. We can, in fact, imagine ourselves to be, for example, both the creature, while he is telling his story, and Frankenstein, while he is telling his story, which includes the creature's telling of his story. The two opposites, the creature and Frankenstein, which can be seen to be unified because the latter created the former and psychologically contains him, are thus embodied in each reader, who is encouraged to unify the opposites within herself. Mary Favret points out that the three main narratives

> present themselves not as successors to one another, but as three versions of the same tale, one comenting upon and responding to the other two.... *Frankenstein* works not to place significance in any one narrative, but to "maintain a presence" [quoting Julia Kristeva] of each within the others.... The monster's story does not exclude the story of his maker, nor does Walton's tale displace the other two. Rather, the voices intersect, which causes them to create new utterances. We can picture the novel, therefore, as a common plane upon which many stories and many languages intersect, regardless of internal contradiction.[2]

Beth Newman places *Frankenstein*'s framing techniques in the context of discussions by Jacques Derrida and Shoshona Felman on the natures of frames. She explains that narrative framing points outside the tale even while it creates borders:

> Contradictory though this may seem, there is really nothing surprising about it; we are confronting a fresh instance of framing's double logic, the tendency of the frame simultaneously to establish boundaries and to announce, even to invite, their violation.[3]

Just as Cusanus, Pope, and Coleridge emphasize the two major opposing forces in the universe—the expanding and contracting or the centrifugal and centripetal—so *Frankenstein* encourages us to think in huge circles of opposing forces. One force tries to keep the circles firm and true, closed and selfish; the opposing force tries to break out of the circles, to escape the claustrophobia. The centripetal force in this novel, the selfishness of inescapable circles and self-absorption, is the more obvious one to every reader. Less obvious, but just as important, is the centrifugal force, the outward driving force, the force of selflessness which emerges most strongly at the very beginning of the novel in the story of the noble fellow, then again less strongly in the middle with Felix's rescue of Safie's father, and then again, more feebly still, at the end of the novel in Walton's decision to obey the will of his men and turn back from his quest.

There are many possible effects of Mary Shelley's structuring devices. For example, we enter gradually into the bizarre and impossible world of the creature. Beginning with a conventional epistolary style, the novel places us safe in England with Mrs. Saville while we gradually learn of Walton's ambitions. At each level of the narration we become more distant from the everyday world of England, both geographically and spiritually. These levels of decompression allow us to begin the creature's tale with some familiarity and expectation, with less likelihood of automatic rejection.

> The frames ... mark the exclusion of Mrs. Saville—and the reader as well—from the horror of the narratives they contain, and signal an immunity from the seductiveness of the voices that first utter them. At the same time, each frame that we pass through as we read makes the matter at the center seem more highly charged, more significant, more invested with power.[4]

Of course, the levels of narration can be played off against each other. In particular we are surprised when the creature speaks with surpassing eloquence and extreme sensitivity after Frankenstein has painted him a cruel murderer. On the other hand, since we have already seen some of Frankenstein's weaknesses, including his egotism and his self-imposed isolation, we now find stronger reasons to question Frankenstein's claims to benevolence when the creature tells us how his creator deserted him. Mary Favret infers from the "multiple, competing voices in the structure of *Frankenstein*," that "like Frankenstein's monster, the novel itself is a representation of human life which exceeds the dimensions of any one individual."[5]

An even more complex relation among narratives occurs when we remember Mary Shelley's description of the creation scene as it first arose in her imagination. She seems to assume that Frankenstein would naturally recoil from his creature:

> I [Mary Shelley] saw the hideous phantasm of a man ... show signs of life.... Frightful must it be.... His success would terrify the artist; he would rush away from his odious handiwork, horror-stricken. He would hope that, left to itself, the slight spark of life which he had communicated would fade. (9)

Thus it is not so easy, as most readers naturally do, to blame Victor Frankenstein for his rejection of the creature when Mary Shelley so easily assumes such a rejection. Furthermore, Frankenstein has told Walton to beware of the creature's speaking ability; the creature may deceive Walton and us. We must weigh the murders committed by the creature against his claims that he is another Adam who asks only for humane treatment.[6]

Finally, because of these kinds of complications among narrative levels we must make judgments as to the worth and reliability of their claims. Where should we place our trust? As readers of this novel, how do we interpret narrative technique to help us decide what values, if any, are being urged? To put my question in other terms, how does structure imply theme?

I would like to focus my ideas about structure and theme on two key episodes in *Frankenstein*. One episode, in the center of the novel, is the most despairing of all; the other, near the beginning, is the most hopeful. I suggest that this combination of hope and despair, a coincidence of opposites, structures the entire novel. In fact, the entire novel and these two episodes in particular invite us to go out of ourselves and imagine behavior which seems impossible for us. They invite us to become wholly other in a way that anticipates the radical self-annihilation in favor of another that Blake will explore and insist upon.

Joseph Kestner explains that not only the content, but also the structure of *Frankenstein* is peculiarly narcissistic. Following Jean Ricadou, Kestner explains that the technique of a story within a story, "an enclosed narrative challenging the primary narrative, becomes structurally a narcissistic text."[7] However, competing with that narcissism, I add, opposing it thematically and structurally, is selflessness.

Percy Bysshe Shelley's definition of morality in his "Defence of Poetry" provides a useful explanation of the effect of the novel and particularly of the two episodes on which I will focus:

> The great secret of morals is Love; or a going out of our own nature, and an identification of ourselves with the beautiful which exists in thought, action, or person, not our own. A man, to be greatly good, must imagine intensely and comprehensively; he must put himself in the place of another and of many others; the pains and pleasures of his species must become his own. The great instrument of moral good is

> the imagination; and poetry administers to the effect by acting upon the cause. Poetry enlarges the circumference of the imagination by replenishing it with thoughts of ever new delight.[8]

Although Shelley's image here does not go as far in becoming other as we shall see Blake's images doing, he does define consciousness and sympathy as a circumference instead of a center. His definition still implies the ego acting from a center, but his centrifugal language, his throwing of the self out into another and his attention to circumference rather than center, not only anticipate Blake but also provide a way to understand the concentric circles of *Frankenstein*. Since the novel is constructed of concentric circles, and since the largest concentric circle by implication is that of the reader's understanding, we are invited to consider ourselves as expansively containing Mary Shelley, Walton, Victor Frankenstein, the creature, and the De Laceys. We thus can become a circumference of sympathetic understanding instead of a centered ego separate from others; we can become universal much like Blake's Albion.

Percy Shelley's definition of morality, similar to that of Hazlitt in his *Principles of Human Action*, goes to the heart of the question of the utility of literature. Instead of a narrow didacticism, Percy Shelley's definition and Mary Shelley's novel give us an enlarging of the imagination, not only as we imagine something new to us, but as we hold mutually exclusive ideas in the mind at the same time. Considered together, the two incidents on which I will focus bring together in a coincidence of opposites the hope and the despair that are so powerful in the novel *Frankenstein*. At the same time each episode on its own invites us to become like the characters: able to identify so fully with another that we lose our own sense of self-preservation.

The first key episode occurs at the structural center of the novel, only a few pages from the physical center. In fact if we include the title page and the dedication page, we find this episode on the 109th page of 218 pages in the Oxford paperback. If we attempt to peer through the layers within frames, to penetrate the circles of narration, the stories within stories, if we search for the center by mathematical or psychological or thematic means, we find this scene. It takes place within the creature's telling of his own life and that of the De Laceys. As he observes the family in their cottage, he is learning a few words, including the contraries "Felix" and "unhappy." Each night he benevolently supplies the cottagers with wood. He hopes that after he learns to speak, the cottagers will accept him as a friend:

> I easily perceived that, although I eagerly longed to discover myself to the cottagers, I ought not to make the attempt until I had first become master of their language; which knowledge might enable me to make them overlook the deformity of my figure. (113–14)

But in the middle of this most hopeful part of the creature's existence, during the arrival of his first spring, he sees his own reflection in a pool of water:

> I had admired the perfect forms of my cottagers—their grace, beauty, and delicate complexions: but how was I terrified, when I viewed myself in a transparent pool! At first I started back, unable to believe that it was indeed I who was reflected in the mirror; and when I became fully convinced that I was in reality the monster that I am, I was filled with bitterest sensations of despondence and mortification. (114)

This moment is the most pessimistic moment in a very pessimistic book. As Harold Bloom explains, "All Romantic horrors are diseases of excessive consciousness, of the self unable to bear the self."[9] Or, to follow Peter Brooks, we encounter "the pathos of a monsterism in doomed dialectic with nature and with culture."[10] This scene blasts any possibility that nurture could ever be the salvation of the creature. Already we have seen him rejected by his creator, by an innocent child, and by others; now we see him rejected by himself.[11] This scene echoes the scene already quoted when Mary Shelley simply assumes that the creature must be rejected by his creator. Like Victor Frankenstein and like the young boy, who both recoiled in horror, the creature, without any training in society's standards, rejects the ugliness of the creature himself. He naturally prefers the beautiful cottagers. His existence is irredeemably rotten to the core. Despite his intelligence, grace, and insight, the creature is doomed not only to inferiority, but to total rejection, total self-hatred. He stands rejected not only by his two creators, Victor Frankenstein and Mary Shelley, neither one of whom gives him a name, but also by himself. How then could he ever be assimilated into any society of humanity?

Kestner points out that a reflection "paradoxically presents a surface of depth"[12]: the gazer can believe that he is seeing deeply, but of course he is seeing only the surface. Thus the creature falls into the same mistake that all observers of him fall into: they think that his superficial surface is really the depth of his being. As readers of the novel, who never see the creature, nor hear him, nor feel him, we are able to bypass the superficial appearance and see the soul of the creature in a way that no character in the book can. Peter Brooks borrows Lacanian terms to explain the difference between the "speculary" mirror-stage and the "symbolic" use of language. The "hopelessness of speculary relationship" when people see him is combatted by the symbols of language, which he learns from watching the De Laceys and which he hopes will be his means of intercourse with others. But the speculary wins out over the symbolic.[13]

Mary Shelley's parody of a central scene from Milton's *Paradise Lost* underlines the poignancy and hopelessness of the creature. After Eve awakens from her creation, she catches a glimpse of her reflection in a lake and is so delighted that she returns for a second look:

> That day I oft remember, when from sleep
> I first awak't, and found myself repos'd
> Under a shade on flow'rs, much wond'ring where
> And what I was, whence thither brought, and how.
> Not distant far from thence a murmuring sound
> Of waters issu'd from a Cave and spread
> Into a liquid Plain, then stood unmov'd
> Pure as th' expanse of Heav'n; I thither went
> With unexperienc't thought, and laid me down
> On the green bank, to look into the clear
> Smooth Lake, that to me seem'd another Sky.
> As I bent down to look, just opposite,
> A Shape within the wat'ry gleam appear'd
> Bending to look on me, I started back,
> It started back, but pleas'd I soon return'd,
> Pleas'd it return'd as soon with answering looks
> Of sympathy and love; there I had fixt

> Mine eyes till now, and pin'd with vain desire,
> Had not a voice ... warn'd me ... (IV.449–68)

This narcissistic display of vanity foreshadows Eve's later susceptibility to Satan's temptations. She seems congenitally incapable of standing firm. But beyond this obvious susceptibility lies a Miltonic sense of self-respect. Eve's beauty, although it may lead to dangerous narcissism, is primarily evidence of God's goodness. Any temptation that the human condition is prey to, whether from physical beauty or from free will, must be withstood, according to Milton. Responsibility remains, no matter what, whether in Eden, or in the wide world at the end, when Adam and Eve carry "Paradise within." Eve's self-love can be either a foundation of self-respect or a temptation to egotistical narcissicism. Similarly, Adam's adoration of Eve can be either self-destructive uxoriousness or generous unity; indeed, when Adam makes the noble decision to fall with Eve, his action can be interpreted in either way, just as the fall itself can be seen as fortunate when viewed as a *felix culpa*.

No matter how a reader may interpret events in *Paradise Lost*, one thing is certain: existence itself, creation of humanity is a blessing, never to be rejected. During Adam's most desperate attempt to defend his sin, he demands of God:

> Did I request thee, Make, from my clay
> To mould Me man? Did I solicit thee
> From darkness to promote me?— (X.743–745)

This defensive blame of God is not allowed to stand within the context of the whole poem. Adam even begins to refute himself a few lines further on:

> ... inexplicable
> Thy Justice seems; yet to say truth, too late
> I thus contest; then should have been refus'd
> Those terms whatever, when they were propos'd:
> Thou didst accept them; wilt thou enjoy the good,
> Then cavil the conditions? (754–759)

Milton's justification of the ways of God to man is rewritten by Mary Shelley as a natural and justified resentment toward a bungling and uncaring manufacturer. Quite clearly in *Paradise Lost* Adam has no right, according to Milton, to question God, to question God's creation, to prefer non-existence. Just as clearly in the novel *Frankenstein*, the creature has every right to question his creator. The creature has been naturally rejected by his creator, by all other humans he meets, and even by himself; the creature naturally and justly rejects his creator, all other humans, and even himself. Not only the creature's ugliness, not only his self-respect, but even his very existence and the very existence of humanity are called into question. The bleak vision could hardly be more nihilistic. The constructive, persistent tone of *Paradise Lost* has been reversed in the mirror of *Frankenstein*'s destructive, dismal resignation.

The tale of the central story of the De Laceys, the one in which this pessimistic mirror scene is enveloped, ends when they also reject the creature. And this rejection takes the pessimism to yet a deeper level. Many readers misinterpret the scene of Father De Lacey's rejection of the creature. Most seem to think that De Lacey accepts the creature until his sighted family returns and chases away the hideous looking monster. But a closer reading reveals that the old man has already rejected the creature before the family sees him, because of the very nature of the desperate creature. Hoping that he can escape the horrors inevitable when anyone sees him, the creature talks to old blind Father De Lacey, who assures him that, as he hopes, "the hearts of men, when unprejudiced by any obvious self-interest, are full of brotherly love and charity" (134). The creature responds with, "where they ought to see a feeling and kind friend, they behold only a detestable monster" (134).

The creature's eloquence has apparently won over the old man, but when the old man's kindness encourages the creature to tell his story and even asks who the creature's potential friends are, the creature panics:

> I paused. This, I thought, was the moment of decision, which was to rob me of, or bestow happiness on me for ever. I struggled vainly for firmness sufficient to answer him, but the effort destroyed all my remaining strength; I sank on the chair, and sobbed aloud. (135)

Thus, even before the rest of the De Lacey family arrives, the creature has already given up. Defining his situation as all or nothing, he makes it into a worse crisis than it might otherwise be. Investing all his energy, all his hopes for the future, in this one event, he can not act. This freezing of his will is an even more excruciating turn of events than his self-rejection in the mirror of the water. For now, even when no one is looking at him, he can not forget himself enough to continue his eloquent plea to Father De Lacey. And Father De Lacey's sensitive hearing probably begins to fear the loud crying.

In his desperate condition, the creature hears the family approach and panics even further, losing the kindness of the old man by his violence:

> At that moment I heard the steps of my younger protectors. I had not a moment to lose; but, seizing the hand of the old man I cried, "Now is the time!— save and protect me! You and your family are the friends whom I seek. Do not you desert me in the hour of trial!"
>
> "Great God!" exclaimed the old man, "who are you?" (135)

The old man, feeling the strength of the creature's grip and hearing the desperation of his voice, recoils before Felix rushes in and drives the creature away. The old man's sense of hearing and his sense of touch have been assaulted just as terrifyingly as has the sense of sight of other characters. Even if the family did not arrive at this time in the action, the old man would probably reject the fierceness of the creature.

Among the readers who remember this incident more benignly—thinking that Felix sees the horrible creature and chases him away just as the blind father might be about to accept him as a friend—is Anne Mellor:

> Whether the blind Father De Lacey reads the Creature's innate character correctly, we as readers can never know, because he is ripped out of the novel by his prejudging son.[14]

Similarly Mary Poovey blames the rejection on Felix and Safie:

> Although the monster tries to disguise its true nature by confronting only the blind old father, De Lacey's children return and recognize the creature's 'ineffaceable' monstrosity for what it literally is.[15]

On the contrary, the blind father rejects the creature well *before* Felix arrives because the creature terrifies him with the strength of his grip and the desperation of his sobbing. Sighted people who meet the creature do not need to ask "Who are you?" because they can clearly see and reject. The blind man needs to ask because he can not give a name to the horror inspired by the grip and the voice. Deeper, then, than the rejections by creator, society, and self based on physical appearance, is this rejection based on a fundamental disposition. The creature is unable to associate with humans because his own self-image, like his reflection, is so damaged, so incapable of constructive use, so irredeemably isolated that he must panic at every possibility of human intercourse. He will be unable even to associate with the benevolent blind because of his self-hatred.

On the next day the cottagers move out, and the creature in his desperate anger burns the cottage and the hovel. In the words of Richard J. Dunn,

> At the very heart of *Frankenstein*, in the tale told by the one narrator who attempts to reach inward and connect himself intimately with the story he tells, we have the motivating repulsion that produces the book's most terrifying events.[16]

Dunn is emphasizing the rejection by the De Laceys, but I find the creature's rejection of his own reflection even more terrifying because it contains, implies, and objectifies his rejection by the old man. Although Felix may recoil from the creature's appearance, Father De Lacey has already rejected the handgrip and the tone of voice. Knowing he is so ugly makes the creature incapable of human intercourse.

Together all the rejections in the novel create a fundamental pessimism that seems to be an absolute rejection on Shelley's part of the idealisms of her father and her husband. Contrary to the theories of William Godwin, education and knowledge do not lead to happiness. Contrary to the *Prometheus Unbound* of Percy Bysshe Shelley, good will can not break the cycle of vengeance.

However, the second scene I would like to focus on works in the opposite direction as a simultaneous contrary. Combined with the central pessimistic scene, it works powerfully as an opposite. This scene takes place near the beginning of the novel. In Walton's second letter to his sister he tells her the story of an Englishman, who is the master of the ship.

> Some years ago, he loved a young Russian lady, of moderate fortune; and having amassed a considerable sum in prize-money, the father of the girl consented to the match. He saw his mistress once before the destined ceremony; but she was bathed in tears, and, throwing herself at his feet, entreated him to spare her, confessing at the same time that she loved another, but that he was poor, and that her father would never consent to the union. My generous friend reassured the suppliant, and, on being informed of the name of her lover, instantly abandoned his pursuit. He had already bought a farm with his money, on which he had designed to pass the remainder of life; but he bestowed the whole on his rival, together with the remains of his prize-money to purchase stock, and then himself solicited the young woman's father to consent to her marriage. But the old man decidedly refused, thinking himself bound in honour to my friend; who, when he found the father inexorable, quitted his country, nor returned until he heard that his former mistress was married according to her inclinations. "What a noble fellow!" you will exclaim. (20–21)[17]

This episode is just as structurally important as the central story within a story which tells of the creature's rejection of his own reflection because it is the first story about someone else in Walton's letters. Thus it is the first story within a story; as such it anticipates the structural device of stories within stories that shapes the entire novel. It is a contrary counterpart to the central structural episode. Whereas the central story of the mirror image in the water reveals nihilistic self-rejection, the first story-within-a-story, by contrast, reveals generous self-abnegation. Thematically the tale of the noble fellow contradicts the selfishness of Frankenstein and the despair of the creature by placing positive value in the negation of the self before we even know the rest of the story. In fact, since the story occurs so early in the novel, before we have encountered the horrors to come, we might even begin to think that this selflessness, this disinterest will serve as the theme of the novel.

Walton has already told us that one of his goals is the betterment of mankind:

> you cannot contest the inestimable benefit which I shall confer on all mankind to the last generation, by discovering a passage near the pole to those countries, to reach which at present so many months are requisite; or by ascertaining the secret of the magnet, which, if at all possible, can only be effected by an undertaking such as mine. (16)

Although he also admits or implies many selfish motives, especially desire for glory, his whole project seems based on a willingness to risk his very existence for some greater good outside himself. Thus Walton's character sets up for us in his own generous noble action and in his tale of the noble fellow a theme of self-sacrifice.

Paul Cantor and Michael Moses, in their overview of the novel, conclude that

> The Creature ... ends up speaking for the value of domestic life in opposition to Frankenstein, who, in his heroic quest as a creator, rejects the ties that would bind him to a conventional family. The Creature longs for precisely the warmth of hearth and home that its creator fails to appreciate.[18]

From this point of view, one might say that so powerful is the desire for domesticity on the part of the noble fellow that, rather than let domestic bliss fade away when he can not achieve it, he creates it for others: the one he loves and one whom he does not know.

The two scenes that I have focused on here, one the very center of the stories within stories and the other the first of the stories within stories, provide precisely opposite views of human nature, and the book as a whole sets before us one of the fundamental questions of theology and of Romanticism: is man good or bad? Of course in theology, that question often takes the form of theodicy: if a perfectly good God created us, then why do we lead such a painful existence? In Romanticism, as in *Frankenstein*, the form of the question often is: if we feel within ourselves such potential for perfection, then why do we fall short? Percy Bysshe Shelley, and even more so William Blake, tell us that we can fully succeed in our idealistic visions; Pope and Coleridge tell us that we can not. Where does Mary Shelley appear to stand in *Frankenstein*?

The creature's rejection of his own image, especially when considered in juxtaposition with the defiant epigraph from *Paradise Lost,* implies that human existence starts horribly flawed and has no chance of redemption. On the other hand the tale of the noble fellow implies that selfless actions can redeem human existence; disinterest can create a larger world of morality, expanding the circumference of the imagination.

This largest opposition, as focused in the two scenes I have chosen, is also built up from many smaller examples. Juxtapositions of opposites occur throughout the novel. According to Peter Brooks, the monster's eloquence gives testimony to his mastery of oppositions expressed in language:

> the Monster is eloquent. From his first words, he shows himself to be a supreme rhetorician of his own situation, one who controls the antitheses and oxymorons that express the pathos of his existence.[19]

Oppositions have structured the creature's existence from the very beginning of his self-education. Among his earliest experiences, he encounters one of the most elementary and important oppositions when he discovers fire:

> I found a fire which had been left by some wandering beggars, and was overcome with delight at the warmth I experienced from it. In my joy I thrust my hand into the live embers, but quickly drew it out again with a cry of pain. How strange, I thought, that the same cause should produce such opposite effects! (104)

As so often, the creature reacts with a surprising sophistication. Primitive minds actually accept the coincidence of opposites quite easily. The creature, however, supposes a kind of logic typical of an Aristotelian philosopher: effects should retain the qualities of their causes. Therefore the same cause can not create unlike, let alone opposite, results. Similarly, the creature interprets his first book, Volney's *Ruin of Empires*, which Felix reads out loud, as a lesson in opposites:

> These wonderful narrations inspired me with strange feelings. Was man, indeed, at once so powerful, so virtuous, and magnificent, yet so vicious and base? He appeared at one

> time a mere scion of the evil principle, and at another, as all that can be conceived of noble and godlike. (119)

Thus in his earliest lessons, the creature learns the inextricability of oppositions in fire and in people.

In fact, even the most negative central episodes of the novel, the rejection of the creature by himself and by the De Laceys, contain their contrary positive values. The entire story of the De Laceys originates in an act of disinterested virtue when Felix springs to the defense of a Turk who is a victim of prejudice:

> A Turkish merchant ... was tried, and condemned to death. The injustice of his sentence was very flagrant; all Paris was indignant; and it was judged that his religion and wealth, rather than the crime alleged against him, had been the cause of his condemnation.
>
> Felix had accidentally been present at the trial; his horror and indignation were uncontrollable, when he heard the decision of the court. He made, at that moment, a solemn vow to deliver him, and then looked around for the means. (122)

Felix has no motive other than a search for justice. Although the Turk turns out to be a villain, Felix does fall in love with the Turk's daughter Safie, who joins the De Lacey family while the creature is watching them and learning from his observations. Thus the main lesson in human behavior that the creature learns from the De Laceys is the beauty and even the reward of disinterested benevolence. As readers, we have been prepared for disinterested benevolence by the tale of the noble fellow almost at the beginning of the novel. The creature does not know that story, but he sees the benevolence of the cottagers toward each other. Just as in the tale of the noble fellow, disinterested action for the benefit of others leads to that happiest of endings, the ever-after marriage of two lovers. Furthermore, even though the De Laceys are poor, their virtues shine through, and the arrival of Safie brings them happiness. Safie, wisdom, joins with Felix, happiness. When they leave the story, we can easily believe that their happiness will continue. They escape the claustrophobic enveloping structures of the novel, in a way analogous to the way the female babe escapes from Blake's "The Mental Traveller" (discussed in Chapter 6). This claustrophobia never really threatens the reader, who may briefly identify with troubled characters' plights; but the escape of the De Laceys may help the reader to perceive her privileged outer position and to learn from it.

Mary Favret, in her analysis of *Frankenstein* and Jane Austen's *Persuasion*, suggests that "the desire to make sense of communication and community propels both works." But while in *Persuasion* we find

> a moment of hope that communion will be realized in the novel, in *Frankenstein* that fragile moment is crushed. Instead the novel reaches outside of itself for coherence—to us, its audience and to the "World."[20]

Thus, we as readers, not being trapped in the novel, have the ability to contain the novel and to learn lessons that the characters can not learn.

Even the creature's rejection of his reflection, the most gloomy scene in the novel, is immediately contrasted by its opposite, happy contrary. Immediately after the creature tells of seeing his horrible reflection in the pool, he tells that his generous stockpiling of wood leads the cottagers to call him, the unknown benefactor, "good spirit, wonderful" (115). He has been spontaneously acting with disinterested benevolence. In the novel as a whole the selflessness of the noble fellow, of Felix, and sometimes even of the creature himself contrasts with the despairing destructiveness that results from the egotistical actions of Frankenstein, which in some ways are similar to the egotistical searchings of the primary narrator, Walton.

The most destructive force in the novel could certainly be defined as greedy egotism. Both Walton and Frankenstein turn away from their families in order to pursue their obsessive goals; the destructiveness of the creature can be seen as an effect caused by those egotistical obsessions. Robert Kiely explains the creature's most heinous murder, that of Victor's bride Elizabeth, in terms of egotism:

> The enemy is an egotism which, when carried to the extreme, annihilates all life around it and finally destroys itself.... While the main theme of the novel is the monstrous consequences of egotism, the counter-theme is the virtue of friendship.[21]

From my perspective, an even stronger counter-theme, one of self-abandoning distinterest, works to produce a coincidence of opposites in the novel. Although Kiely's main counter-theme is certainly the most obvious one, noticed by many commentators, I think that an even stronger counter-theme, placed first among the stories-within-stories and then again at the center of the stories-within-stories, provides a selfless force, equal and opposite to the force of egotism. Even the last event in the novel, Walton's decision to turn back, is based on sacrificing egotistical desires for the happiness of others. He is not disinterested and generous, as are the noble fellow and Felix, but he does make a decision opposite to the theme of egotism. He turns back, even though he does not wish to. We encounter then in three key structural episodes—at beginning, middle, and end—powerful rejections of the self, whether in self-abnegation, self-hatred, or self-denial.

Many critics have seen in *Frankenstein* a kind of indeterminacy and have placed Mary Shelley in contexts which shed light on the question as I have defined it. L.J. Swingle uses an epistemological approach to praise *Frankenstein* as a tragedy:

> By means of multiple first-person narration, the balancing of unresolved conflicting claims to truth and justice, and ambiguous primary evidence, Mary Shelley prevents the reader from knowing the monster. By so doing, I believe, she heightens her novel's significance, transforming it from a fairly simple moral tract into something approaching tragedy.[22]

Charles Schug, acknowledging his debts not only to the ideas of Swingle, but also to those of Richard Dunn, Karl Kroeber, and Robert Langbaum, emphasizes reader participation and an on-going moral experience which is similar to that of Romantic poetry:

> because of the manner in which the novel is constructed, the real moral problem of *Frankenstein* is an experiential problem: it cannot be comprehended or even approached outside its embodiment in Frankenstein's, the monster's, and Walton's experiences; it cannot be apprehended outside the reader's direct experience of it.[23]

Both of these opinions are compatible with mine. However, I think that the structural accomplishment of *Frankenstein* can be stated more fundamentally with a little help from Carl Jung and even more help from Percy Bysshe Shelley.

The creature in Mary Shelley's novel has no name. He has no friend. Despite his best intentions he becomes a serial murderer. These characteristics, and many others, anticipate the shadow as explained by Carl Jung. Frankenstein, in his refusal to acknowledge his monster, his shadow, looks forward to the error of twentieth-century man, as defined by Jung:

> We [modern man] ... have repudiated it [the shadow, "the night-side of life," "the nocturnal world"] ... because we strive to construct a conscious world that is safe and manageable.... Yet, even in our midst, the poet now and then catches sight of the figures that people the night-world—the spirits, demons and gods.[24]

The creature repeats his creator's error when he can not face his own reflection. That is, just as Frankenstein must naturally recoil in horror when he sees his dark creation, so the creature naturally recoils in horror when he sees himself. So the old De Lacey naturally rejects the violently desperate handgrip and voice of the creature. So Mary Shelley naturally assumes that Frankenstein must naturally reject his creature. It is natural for us not to want to see our dark, monstrous shadows. Our egotism, which wants certainty and self-esteem, will naturally not allow us to do so. What comes naturally blocks the very actions carried out and thereby implicitly recommended by the disinterested characters. The first law of nature is self-preservation.

However, as one of Jung's students puts it, when the individual finds herself in a spiritual crisis,

> There is only one thing that seems to work; and that is to turn directly toward the approaching darkness without prejudice and totally naively, and to try to find out what its secret aim is and what it wants from you.[25]

Victor Frankenstein of courses refuses to do what his dark side demands: to love or at least to acknowledge him. He even finds excuses not to make a female companion for the creature.

The novel's theme and structure invite us to confront the creature as we read the book and to compare ourselves to the main characters as they make decisions or helplessly fulfill their fates:

> It is often tragic to see how blatantly a man bungles his own life and the lives of others yet remains totally incapable of seeing how much the whole tragedy originates in himself, and how he continually feeds it and keeps it going.... it is quite within the bounds of possibility for a man to recognize the relative evil of his nature, but it is a rare and shattering experience for him to gaze into the face of absolute evil.[26]

The novel shows us how nearly impossible gazing into evil can be. Many a theodicy, like the one of Alexander Pope, explains away the existence of evil. *Frankenstein* shows us characters who turn away from the horrible vision, but the structure of the novel insists that we look directly into the depths, that we forget our safety and egotism as readers. We can easily see how Frankenstein bungles his life and how the creature lives out that bungling, but can we see how this novel reveals to us our own bungling, our own egotism, our own inability to face our shadows? Can we look into the reflecting surface that will reveal our monstrous, dark shadow to us? The De Laceys escape from the center of the novel, and we escape through the layers of narration. Like Mrs. Saville at home in England, we are safe: we read about the creature instead of seeing, hearing, and feeling him with our senses. We are led through the consciousnesses of the different narrators to the heart of darkness and then allowed to ascend unharmed. We can turn our eyes away from the book at any time. But if we have not faced the monster, then we have failed to learn what the structure and theme of this novel can teach us. We have missed the experience they offer.

All theories of human behavior fail before experience, and all attempts to avoid experience lack wisdom and depth. In Percy Bysshe Shelley's dialogue poem, "Julian and Maddalo," the two title characters, thinly disguised versions of Shelley and Byron, debate the cause of insanity and misery. Shelley of course takes the idealist's position, Byron the realist's. They visit a maniac, whom each wants to claim as proof of his theory. But after the visit,

> ... our argument was quite forgot,
> ...
> And we agreed his was some dreadful ill
> ...
> And I remember one remark which then
> Maddalo made. He said: "Most wretched men
> Are cradled into poetry by wrong,
> They learn in suffering what they teach in song.[27]

The power of Romantic poetry, including *Frankenstein*, is to teach in song that which philosophy, theory, the world of light can never teach. The closest that abstract definitions can come is to call the lesson the giving up of the self. The noble ship's master gives up his egotistical desires, and we in the "real world" find his behavior baffling, stupid, ridiculous. We look the other way. He is not natural. Victor Frankenstein acts out his unconscious desires and destroys all he loves. We look the other way. He is not natural. The creature recoils from his own image. We look the other way. He is not natural. But the novel *Frankenstein* invites us to look at these actions, in the characters and in ourselves.

Percy Bysshe Shelley's preface to *Frankenstein*, written in the author's voice, insists,

> I am by no means indifferent to the manner in which whatever moral tendencies exist in the sentiments or characters it contains shall affect the reader; yet my chief concern in this respect has been limited to the avoiding the enervating effects of the novels of the present day, and to the exhibition of the amiableness of domestic affection, and the excellence of universal virtue. (13–14)

Interestingly, Richard Dunn, while writing an excellent article that I have already cited, bases many of his ideas on a misreading of this statement. He misconstrues the syntax so that the two final phrases (starting "to the exhibition") fall into the category of those things which the author is avoiding;[28] clearly, however, Percy Shelley is claiming that *Frankenstein does* exhibit domestic affection and the excellence of universal virtue. He could be wrong about his wife's novel, of course. At the MLA Convention of 1986 and in subsequent publications,[29] Anne Mellor weighed his heavy hand in many passages. I think in this case Percy Shelley is right, and my approach is based on honoring the intention he claims in order to find a substantial and inclusive interpretation of *Frankenstein*.

Mary Shelley shares with her characters Walton and Frankenstein the ambition to surpass the everyday world:

> my favourite pastime ... was to 'write stories.' Still I had a dearer pleasure than this, which was the formation of castles in the air—the indulging in waking dreams—the following up trains of thought, which had for their subject the formation of a succession of imaginary incidents. (5)

Insofar as we enter into this novel, we are also preferring imagination over the limits of the so-called real world. In the real world, the shadow is not confronted; it may not even exist. Only the imagination—with the tool of the *coincidentia oppositorum*—can find the courage and the techniques that allow such confrontation. We enter the world of imagination, we confront our shadows, at great cost, but we thereby take the chance that we might discover the place within ourselves where opposites coincide, where egotism and selflessness, hope and despair come together. The triumph of Romanticism is to take that chance. The triumph of *Frankenstein* is to construct a narrative technique which invites us to plunge in.

Anne Mellor interprets the novel as revealing Mary Shelley's preference for love over egotism, as she distinguishes between the sublime and the beautiful:

> Because the mind is more likely to respond to the sublime or the unknown with fear and hostility than with love and acceptance, the unfettered imagination celebrated by the Romantic poets is more likely to construct evil than good. Mary Shelley believed, I argue, that the Romantic imagaination must be consciously controlled by love, specifically a love that sees all the products of nature—the old, the sick, the handicapped, the freaks—as sacred life-forms to be nurtured with care and compassion.... As Burke wrote, the sublime appeals to the instinct of self-preservation and rouses feelings of terror that result in a desire for power, domination, and continuing control. But the beautiful appeals to the instinct of self-procreation and rouses sensations of erotic and affectional love.[30]

In a similar reading, Betty T. Bennett sees *Frankenstein* as answering *Alastor,* and then *Prometheus Unbound* as answering *Frankenstein* in variations on the theme of "the necessity for a value system based on universal, selfless love."[31]

I agree with those interpretations insofar as they try to penetrate Mary Shelley's desires and motives. However, what Shelley actually gives us in the novel is an excruciating tension between those two forces, rather than a clear preference for one or the other. As does her husband in *The Triumph of Life,* she may be suppressing her own hopes in a presentation of opposites that should but do not coincide.

She tries to find a coincidence of opposites by enclosing the egotistic Victor Frankenstein within the circumference formed by the selflessness of the noble fellow at the beginning and Walton's turning back at the end. In terms of concentric circles, those two denials of the self form the outer circle. She furthermore encloses Victor Frankenstein's egotism by placing the story of the De Laceys at the center of the novel. As the central, inner circle, the beneficent De Laceys, with their story of distinerested benevolence, contain the egotism between themselves and the outer circle. Thus both center and circumference of the novel *Frankenstein* unite in their containing of the selfish ego that so dominates the rest of the novel.

Mary Shelley therefore can be placed in between the conservatism that we find in Cusanus and Pope and the radicalism that we find in Percy Shelley and Blake. While denying, as do Cusanus and Pope, that man can achieve the ideal realm, she nevertheless seems unable to rest content with the realm of reality. If we apply the rubric "All Nature is but Art" to *Frankenstein,* we might conclude that the creature can never be natural in any common sense of the word. All other human beings can assume some kind of natural, even God-given existence, but the creature, whose creator categorically rejects him, must try to create his own existence. He can not do so, not in the novel, not in any conceivable set of circumstances.[32]

When the creature finds that his "natural" being has actually been created by a psychological monster, in a twist on the Popean optimism of "All Nature is But Art," the hidden truth of the artist turns out to be a nightmare instead of a blessing, as in Pope and Blake. Here the "nature" of the creature, which includes his horrible, unbearable appearance, turns out to be the creation of a debased parody of God who never has had the best interests of his creature in mind. Although by "nature" the creature is benevolent and becomes impressively articulate, also by "nature" on a more fundamental level he is completely unacceptable to society and even to his creator, and even to his own self.

But the acts of selflessness in the novel, especially that of the noble fellow, sacrifice the happiness of the selfish ego and create another kind of existence, not at all confined to the limits of the natural. They therefore lean towards the acts of self-sacrifice that we shall see in Percy Shelley and ultimately in Blake's "Self-Annihilation."

Notes

1. Mary Shelley, *Frankenstein*, ed. M.K. Joseph (NY: Oxford UP, 1969), p. 91. All quotations of *Frankenstein* are from this edition.

2. Mary Favret, *Romantic Correspondence* (Cambridge UP, 1993), pp. 182, 186, 188.

3. Beth Newman, "The Frame Structure of *Frankenstein*," *E.L.H.*, 53 (1986), p. 154.

4. Newman, p. 159.

5. Favret, p. 178.

6. L.J. Swingle, "Frankenstein's Monster and Its Romantic Relatives," *Texas Studies in Literature and Language*, 14 (1973), p. 52.

7. Joseph Kestner, "Narcissism as Symptom and Structure," in *The Nature of Identity* (Univ. of Tulsa, 1981), p. 17.

8. *Shelley's Poetry and Prose* (Norton, 1977), pp. 487–88.

9. Harold Bloom, "Introduction," in *Mary Shelley's* Frankenstein, ed. Harold Bloom (Chelsea House, 1987), p. 9.

10. Peter Brooks, "'Godlike Science/Unhallowed Arts,'" in Levine and Knoepflmacher, *The Endurance of* Frankenstein (Berkeley: Univ. of California Press, 1979), p. 205.

11. Although not directly on the subject of the reflection in the water, Alan Richardson's comments on the education of the creature emphasize his "irreducible difference" from humanity. The viewing in the pool clarifies that difference in the mind of the creature. Comparing the education of the creature to that of women in Mary Shelley's day, Richardson, quoting Rousseau, points out that a young woman was trained to restrain her unlimited desires in "a perpetual combat against herself." The creature's natural sympathies are now turned against himself as he realizes that no matter how hard he tries, he can never join the human race. Like the women of the early nineteenth century, according to Richardson, the creature must learn his inferiority and deformity as he internalizes social values. ("From *Emile* to *Frankenstein*," *European Romantic Review*, 1 [1991], pp. 157, 151.)

12. Kestner, p. 18.

13. Brooks, pp. 207–08.

14. Anne K. Mellor, "*Frankenstein* and the Sublime," in Behrendt, *Approaches*, p. 100.

15. Mary Poovey, *The Proper Lady and the Woman Writer* (Chicago: Univ. of Chicago Press, 1984), p. 129.

16. Richard J. Dunn, "Narrative Distance in *Frankenstein*," *Studies in the Novel* (North Texas State University), 6 (1974), p. 414.

17. This story of the noble fellow is very seldom mentioned in *Frankenstein* criticism. For example, it never appears in Stephen Behrendt's recent compilation of essays by several hands, *Approaches to Teaching Shelley's* Frankenstein. Beth Newman, in her analysis of framing effects, does use the story of the noble fellow to emphasize the contrast between the promise kept by the noble fellow and all the broken promises in the novel, including the one at the center by the Turk who refuses his daughter Safie in marriage to Felix (Newman, p. 155).

18. Paul A. Cantor and Michael Valdez Moses, "Teaching *Frankenstein* from the Creature's Perspective," in Behrendt, *Approaches*, p. 131.

19. Brooks, p. 206.

20. Favret, p. 178.

21. Robert Kiely, *The Romantic Novel in England* (Cambridge: Harvard UP, 1972), p. 166.

22. L.J. Swingle, "Frankenstein's Monster and Its Romantic Relatives," *Texas Studies in Literature and Language,* 14 (1973), p. 55.

23. Charles Schug, "The Romantic Form of Mary Shelley's *Frankenstein,*" *Studies in English Literature,* 17 (1977), pp. 612–13.

24. Carl Jung, *Modern Man in Search of a Soul,* trans. W.S. Dell and Cary F. Baynes (NY: Harcourt, 1933), p. 163.

25. M.-F. Von Franz, "The Process of Individuation," in Jung, *Man and His Symbols* (1964; rpt. NY: Dell, 1970), p. 170.

26. Carl Jung, *Aion* (1959; rpt. Princeton UP, 1979), p. 10.

27. Lines 520–46, *Shelley's Poetry and Prose,* p. 125.

28. Dunn, p. 416.

29. Including her "*Frankenstein* and the Sublime," in Behrendt, *Approaches,* pp. 99–104.

30. Mellor, "*Frankenstein,*" p. 104.

31. Betty T. Bennett, "Love and Egocentricity," in Behrendt, *Approaches,* p. 76.

32. The ending of the film *Young Frankenstein* does fulfill the creature's desire for a normal home, but the hulking, urbane, slipper-wearing, newspaper-reading, pipe-smoking Peter Boyle in his easy chair blasts the happy ending with the absurdity of his pose.

Chapter 5
Images and Deep Truth: The Individual Mind and the External World in Four Shelley Poems

Throughout his poetry Percy Bysshe Shelley explores many coincidences of opposites, particularly the identity of the individual mind and the external world. The mind can conceive of ideal unity and can sometimes perceive it in the universe, as in "Hymn to Intellectual Beauty":

> The awful shadow of some unseen Power
> Floats though unseen amongst us,—visiting
> This various world …
> …
> … Like aught that for its grace may be
> Dear, and yet dearer for its mystery.[1]

However, the factual multiplicity of the world, including nature, other people, and even recalcitrant aspects of the individual himself, will not maintain and conform to the ideal vision:

> Love, Hope, and Self-esteem, like clouds depart
> And come, for some uncertain moments lent.
> Man were immortal, and omnipotent,
> Didst thou, unknown and awful as thou art,
> Keep with thy glorious train firm state within his heart.

An idealistic vision, like Shelley's, assumes that the ideal is the source and the end of all things and urges man to work his way through the multiplicity of the real back to the unity of the ideal. This belief is evidenced in several places in Shelley's poetry, especially in *Adonais*. With an opposite emphasis, the empiricist vision denies the ideal in the first place and assumes that man is a *tabula rasa* who learns all he knows from the experience of exterior phenomena. On first glance, these two positions are so different as to be completely irreconcilable. However, Shelley finds a synthesis by seeing no essential difference between the ideas in the mind and the objects in the world:

> The view of life presented by the most refined deductions of the intellectual philosophy, is that of unity. Nothing exists but as it is perceived. The difference is merely nominal between those two classes of thought which are vulgarly distinguished by the names of ideas and of external objects. Pursuing the same thread of reasoning, the existence of distinct individual minds similar to that which is employed in now questioning its own nature, is likewise found to be a delusion. The words, *I, you, they,* are not signs of any actual difference subsisting between the assemblage of thoughts thus indicated, but are merely marks employed to denote the different modifications of the one mind.[2]

Although such a formula sounds idealistic in the extreme, it derives logically from the principles of empiricism: if all we know is what we perceive, then there is no discernible difference between ideas and our perceptions of external objects. Our perceptions of objects exist in our minds as do our ideas; as far as we know, as far as we experience, both ideas and objects exist only in our perceptions. Without our perceptions, we would have no conception of objects, but the existence of those objects outside our perceptions can not be demonstrated. Therefore neither our supposedly internal ideas nor our perceptions of supposed external objects have a superior claim to be denominated internal or external. Both may be external; both may be internal. They may be neither external nor internal; they may be both external and internal. The difference is merely nominal. A fundamental distinction made by everyday common sense meets a coincidence of opposites.

As C.E. Pulos demonstrates throughout his book *The Deep Truth*,[3] the skeptical tradition provides a philosophical framework which contains such a difficulty. The skeptic does not know with absolute certainty that objects exist outside his own mind, but he does know that if they do exist, then his only knowledge of them is in his mind through his perceptions. Therefore, as far as man can tell, there is no way to distinguish between the object and his idea of it. Merle Rubin makes a direct link between skepticism and Shelley's urges toward a coincidence of opposites. Like the forces so crucial to the theories of *coincidentia oppositorum* in Cusanus and Coleridge, the opposing centrifugal and centripetal forces become juxtaposed:

> Another quality of Shelley's poetry that may or may not be attributable to skepticism, but is consonant with that outlook is the sensitivity it displays to the differing beauties of two opposing forces, one centripetal, the other centrifugal. His is a temperament that deeply perceives separateness, discreteness, the flux of the mutable world, the lack of connection between effect and cause, and all manner of incongruity, but which at the same time craves continuity.... Much of his poetry, then, deals with the tension that results from juxtaposing these two divergent modes of perception.[4]

Shelley himself even uses the mystical formula from geometry that Cusanus used to express the unity of the universe based on these two opposing forces:

> This is the character of all life and being.—Each is at once the centre and the circumference; the point to which all things are referred, and the line in which all things are contained.[5]

However, as helpful as this train of thought may be in understanding one paradox, it intensifies the more fundamental one: if idea and object are indistinguishable, then why are they so often in conflict? Why does the world of objects not simply coincide with the world of ideas in a more obvious way? Why are desire and the objects of desire so often separate?

The Coleridgean theory of organicism, as explained by M.H. Abrams in *The Mirror and The Lamp*, finds another way to view the individual mind and the external world, but leaves a similar paradox behind. "Grounded on Schelling's metaphysics of a psycho-natural parallelism, according to which the essences within nature have a kind of duplicate subsistence as ideas in the mind,"[6] Coleridge's view sees the individual creative mind as a kind of plant which grows as naturally as do real plants in nature. Because of the essential similarities between the workings of the mind and the workings of nature, they can be used as images of each other: "to make the external internal, the internal external, to make nature thought, and thought nature,—this is the mystery of genius in the Fine Arts."[7] But if the relation of mind to object is so natural, the question remains: why the conflict? If inner and outer are one, then why does inner usually feel at frustrating war with outer?

Of course certain forms of Christianity, and certain other religious views, provide another traditional answer to the problem: God sees the whole picture and man should trust Him. Man's discontent with the world of objects arises from the smallness of his mind. If man could see more than he can see, a hypothesis which is by definition impossible, then he could understand the apparent disparities between inner and outer. Meanwhile submit to God's infinite will: "Whatever is, is right." This, of course, was Alexander Pope's answer, and in a way it was the answer of Nicolas Cusanus also. Great mysteries abound; God understands. Quiet your anxieties. Refusing to submit to such a God of mystery and implied tyranny, Shelley nevertheless retained his belief in the transcendence of a pervading Spirit. Admitting man's weaknesses, he still insisted on man's essential dignity and integrity.

Four of Shelley's poems will serve as examples of the way he explores the issue of inner and outer in his imagery and of the way he poses questions beyond imagery in order to try to understand the identity of individual mind with the external world. In *Alastor* the hero hastens to his death because he refuses to acknowledge the existence of the outer world in his frantic inner search for the ideal female whom he has constructed in his fantasy. In "Mont Blanc" the poet's individual mind assumes a Power behind all things and poses the question of their relationship in its sharpest terms. In *Prometheus Unbound* Shelley presents a future vision of perfect harmony between the real and the ideal. In *The Triumph of Life* the facts of life crush the idealist and almost everybody else but do not destroy the hope that always exists in Shelley's poetry.

As in Cusanus, Pope, Coleridge, and Blake, much of the power of Shelley's vision arises from a powerful structure of oppositions.

Unlike Blake, but like the others, that structure of oppositions can be, at least partly, propelled by skepticism. Not being able to know for sure which side to choose in a particular issue can lead to oscillation, or suspension of judgment, or in Shelley's case, to a leap of faith. Merle Rubin uses Pulos's idea of skepticism to explain how hope counters despair. If skepticism is a tool with which to doubt idealisms, then it is an equally powerful tool with which to doubt realisms:

> Shelley's skepticism, so purgative a power against idealizing excesses, against the heart's insistence upon continued hope, is also an outlook which can temper the tendency to de-idealize, the head's conviction of despair. For if vision may be doubted, so may reality be questioned.... Skepticism, an intellectual habit that can in some circumstances be paralyzing, is for Shelley (as it was in a way for Hume) a double-edged weapon that can free him to act, as is suggested in the lines from Adonais: 'And keep thy heart light lest it make thee sink / When hope has kindled hope, and lured to the brink.'[8]

In each of these four poems, images of seeing play a central role. In a fascinating implicit footnote to his wife's images of sight in *Frankenstein* (and of course her novel can be read as a comment on his ideas), Shelley shows us that mirror imaging leads to stasis, entrapment, and even death. Like Frankenstein's Creature, who can see only horror in his own reflected appearance, Shelley's personae see only dead ends when they see only themselves. In *Alastor* the poet never finds his way out of seeing only himself. Thus Shelley explores one of the dangers of conceiving of inner and outer as unified: inner may take precedence and subsume outer. If inner and outer are considered the same, then one or the other may be ignored. Seeing through apparent surfaces instead of simply seeing our own reflections can reveal deeper connections and let us break through the apparent limits that stand before us. If outer and inner are only nominal distinctions, then a full vision will incorporate both fully. The questions in "Mont Blanc" lead to such a full vision in *Prometheus Unbound*.

Pope's concept "All Nature is but Art" begins to take on the ideology-blasting force that Blake will give it. Unlike Pope, who leaves mysteries completely intact; unlike Coleridge, whose over-arching ambitions toward wholeness leave him in disappointment; unlike Mary Shelley, who presents the failure to achieve a coincidence of opposites in a context of despair; Percy Bysshe Shelley sets up an elaborate series of mirrors and veils in "Alastor" and "Mont Blanc," which he then pierces in *Prometheus Unbound*. After the triumph of vision in *Prometheus Unbound*, Shelley explores the problems on a deeper and more recalcitrant level in *The Triumph of Life*.

Alastor is full of images that turn in on themselves, mirror themselves, reflect themselves, in various levels, combinations, and repetitions. The poem begins with a contrast between the narrator, who may be Shelley himself, and the poet-hero, who may be a representation of the younger Shelley. The narrator's prayer assumes a reciprocal relationship with nature:

> Earth, ocean, air, beloved brotherhood!
> If our great Mother has imbued my soul
> With aught of natural piety to feel
> Your love, and recompense the boon with mine
> ...
>
> (1–4)

The narrator thus sets up a process of continual interchange in which his poetry will be a repayment for what he has received. He feels that his inner virtue has been given to him from outside, and that he therefore must reciprocally give back in his poetry and praise. The presence or absence of reciprocity becomes a key theme in the poem.

The *Alastor* poet, on the other hand,

> ... lived, he died, he sung, in solitude.
>
> (60)

His isolation is so great that even after death,

> ... no mourning maiden decked
> With weeping flowers, or votive cypress wreath,
> The lone couch of his everlasting sleep.
>
> (55–57)

Paradoxically, the assumption of solipsism, based on an idea of the unity of the self, leads to separation and extinction; awareness of others, based on an idea of the importance of the existence of others and of reciprocity, leads to a finding of the unity which was lost. This theme, implicit in *Frankenstein*, becomes more explicit in the poetry of P.B. Shelley. Although both narrator and poet were brought up in settings of natural beauty, the narrator feels an urge to "recompense the boon" which nature has bestowed on him. The poet seems to have no such urge to reciprocate. Instead of interacting with a nature that has some existence apart from him, he ignores it or imposes his own view on it until all of nature becomes a mere reflection of his glory. But once it becomes a mere reflection, then the poet has destroyed not nature, but himself.

We notice this lack of vision most obviously when he completely ignores the Arab maiden who serves him food. We are told this story right after being informed that "he saw / The thrilling secrets of the birth of time":

> Meanwhile an Arab maiden brought his food,
> Her daily portion, from her father's tent,
> And spread her matting for his couch, and stole
> From duties and repose to tend his steps:—
> Enamoured, yet not daring for deep awe
> To speak her love:—and watched his nightly sleep,
> Sleepless herself, to gaze upon his lips
> Parted in slumber, whence the regular breath
> Of innocent dreams arose: then, when red morn
> Made paler the pale moon, to her cold home
> Wildered, and wan, and panting, she returned.
> (129–139)

The poet's response to this woman's attention is sleep. Instead of a real other, he prefers to dream of an ideal woman. He is so busy traveling and discovering the secrets of the ages that he forgets that he is human, and so instead of loving the Arab maiden, he creates out of his own imagination the ideal female. Of course she is not a real woman who exists outside the poet, but merely a reflection of his own self. He creates her as if he were looking in a mirror in which he saw his own reflection wearing a veil. She is partially hidden from him because he is projecting a part of himself, with which he can fall in love, like Narcissus to a higher power:

> ... A vision on his sleep
> There came, a dream of hopes that never yet

Had flushed his cheek. He dreamed a veiled maid
Sate near him, talking in low solemn tones.
Her voice was like the voice of his own soul
Heard in the calm of thought ...
...
Knowledge and truth and virtue were her theme,
And lofty hopes of divine liberty,
Thoughts the most dear to him, and poesy,
Herself a poet. (149–61)

This projection of the self is even more severe than in a similar scene in Byron's *Manfred*. At least in *Manfred* the beloved, who is also just like the lover, with the same lineaments (and in that case, it is implied, at least one parent in common), except that she is feminine, turns out to be superior in virtue to the dreamer:

She was like me in lineaments—her eyes
Her hair, her features, all, to the very tone
Even of her voice, they said were like to mine;
But softened all, and tempered into beauty
...
Her faults were mine—her virtues were her own
(*Manfred* II.ii.105–116)[9]

The *Alastor* poet's fanciful beloved does not even receive that honor of being morally superior to her creator; she is simply identical to the poet, except that she is veiled and feminine.

After the poet makes love with this creation of his solipsistic imagination, he wakes to a feeling of loss similar to that of Wordsworth in the *Intimations Ode*. Wordsworth writes:

There was a time when meadow, grove, and stream,
The earth, and every common sight,
To me did seem
Apparelled in celestial light,
The glory and the freshness of a dream.
...
...
Whither is fled the visionary gleam?
Where is it now, the glory and the dream?[10]

Shelley echoes:

> Whither have fled
> The hues of heaven that canopied his bower
> Of yesternight? The sounds that soothed his sleep,
> The mystery and the majesty of Earth,
> The joy, the exultation? His wan eyes
> Gaze on the empty scene as vacantly
> As ocean's moon looks on the moon in heaven.
> (196–202)

Ironically the poet bemoans the loss of a glory which arose from his own mind. Like Coleridge when he tries to respond to Wordsworth's *Intimations Ode* (see discussion in Chapter 3), Shelley's poet has set up for himself a short-circuit: with no other being to connect to, he can experience only loss of power, desire, satisfaction, and accomplishment. The veiled maid is a reflection of his fancy, just as the moon in the ocean is a reflection of the moon in heaven. And even the moon in heaven shines only as a reflection of the light of the sun, and so the poet's eyes here are being likened to the reflection of a reflection. Having chosen inner reflection over outer perception, he has lost the joys of both and is left with the torments of solipsism.

All his reflecting leaves the poet in the center of his own thoughts and projections, wondering how to tell the difference between reflections and deeper truths:

> Does the dark gate of death
> Conduct to thy mysterious paradise,
> O Sleep? Does the bright arch of rainbow clouds,
> And pendent mountains seen in the calm lake,
> Lead only to a black and watery depth,
> While death's blue vault ...
> Conducts, O Sleep, to thy delightful realms?
> (211–16, 219)

Once again re-writing Wordsworth, Shelley's poet sees in nature not a redeeming, chastising outer construct with which to engage in reciprocity, but only darkness and a dead-end. Perhaps, he muses, death might be a good idea. Not until *Prometheus Unbound* (see discussion below) does Shelley, in the character of Asia, give any way of seeing through these endless reflections. Instead, hoping to recapture the delightful sleep in which he first encountered the veiled maid, the poet assumes that death might be a permanent sleep. Caught in the problem of lost vision, he thinks that the only solution is to narrow his vision even more strictly.

Hastening toward his death, the poet exists in a self-enclosed system.

> ... At night the passion came,
> Like the fierce fiend of a distempered dream,
> And shook him from his rest, and led him forth
> Into the darkness.—As an eagle grasped
> In folds of the green serpent, feels her breast
> Burn with the poison. (224–29)

Since the passion-serpent which grasps him is a mirror-image creation of his own dreaming mind, as was the maiden whom he loved, he is both eagle and serpent, tortured by himself, flying toward his own image of himself. Later the grasping in the bird-serpent image is reversed: the waves on which the poet's boat floats are "Like serpents struggling in a vulture's grasp" (325). This entanglement of solipsistic passion is echoed by the repetitions of words:

> Calm and rejoicing in the fearful war
> Of *wave* running on *wave*, and *blast* on *blast*
> Descending, and black flood on whirlpool driven
> With dark obliterating course, he sat. (326–29)

The poet delights in seeing the ocean doing the same thing that he is doing: turning in on itself, involving itself in its own passion and enjoying the violent self-destructiveness. In this way, the mirroring repetition is doubled again: nature becomes not a separate exterior but another reflection of the poet. In fact the poet seeks out environments which correspond to this state of mind, and possibly even half-creates them. Thus, while he is caught in the grips of self-reflection, he searches for other examples of self-reflection, thus making even his supposed perceptions of the outside world into mirror reflections of himself.

Shelley's idea stated earlier—that the difference between ideas and objects is merely nominal—is thus pushed to its ultimate extreme. The impulse urges him to places which suit his fancy. It is as if the outside world exists only to satisfy his solipsistic urges, and therefore as if the outside world does not exist at all. The ultimate nightmare of the mis-use of the idealistic vision has been reached.

Into the depths of nature travels the poet: "The boat pursued / The windings of the cavern" (369–70). And yet the windings of the cavern seem not to be nature at all, but only the convolutions of the poet's mind. He arrives at a vortex and finds another image of himself:

> I' the midst was left,
> Reflecting, yet distorting every cloud,
> A pool of treacherous and tremendous calm.
> (384–86)

Just as the pool distorts the things it reflects, so the poet misjudges the world in which he lives and recreates it in his own image. The word "midst" shows his movement toward the middle, a center that is a point of ego isolation. He has once again found or created a more perfect and stultifying analogue to his distorting solipsism.

As the stream winds among mossy banks, the poet sees the narcissi engaging in the same solipsistic self-admiring self-reflection that he engages in. Now not only is he seeing reflections of his own condition, but he is seeing parts of Nature that have been named after the myth of the most intense self-gazer of all—Narcissus. The flowers seem to be engaging in the exact same act as the poet. At some point of course, as in the dynamics of Blake's "The Mental Traveller" (see discussion in following chapter), an extreme intensification of any system or attitude invites its own reversal as it reaches its limits. Indeed at this point the poet has a chance, even a desire, to see something beyond himself, to enjoy nature instead of just using it:

> yellow flowers
> For ever gaze on their own drooping eyes,
> Reflected in the crystal calm. The wave
> Of the boat's motion marred their pensive task,
> Which nought but vagrant bird, or wanton wind,
> Or falling spear-grass, or their own decay
> Had e'er disturbed before. The Poet longed
> To deck with their bright hues his withered hair,
> But on his heart its solitude returned,
> And he forbore. (406–09, 412–15)

Even as the poet continues on his mistaken journey, the boat in which he travels disturbs the calm reflection of the narcissi. The outside world intervenes. But for a completely inexplicable reason, similar to the mysterious reason for Satan's inescapable recalcitrance in Book IV of *Paradise Lost,* the poet seems incapable of escaping the condition of his heart. If the poet could only disturb his own self-involved reflections, perhaps he could, like the Ancient Mariner, bless nature unaware. But he resists his own best impulse.

In the very middle of the forest he comes to a well that

> Images all the woven boughs above,
> And each depending leaf, and every speck
> Of azure sky, darting between their chasms.
> (459–61)

As when he saw the narcissi, the poet could learn a lesson that could shake him out of his self-destructive solipsism; as the well reflects all of nature, so he could notice the rich complexities of the world around him. But instead, when he gazes into the well, he sees only himself, or more particularly, only his eyes, or more solipsistically, only the light of his eyes, the projection of his gaze:

> His eyes beheld
> Their own wan light through the reflected lines
> Of his thin hair, distinct in the dark depth
> Of that still fountain; as the human heart,
> Gazing in dreams over the gloomy grave,
> Sees its own treacherous likeness there.
> (469–74)

Even the poet's own hair has become just another surface for reflections. Like the earlier image that compared the poet's eyes to the moon, this image presents only reflections of reflections. Because he has allowed himself to see only himself, because the only world he wants is a dream world, because the only possible way to achieve his ideal is through death, everything the poet sees turns into an image of death. Death is the end of human life, and constant reflection is the end of his perception.

Just before he dies, the poet seems to achieve some possibility of self-understanding as he poses a central question. Although he still compares everything to himself, still measures existence by the yardstick of his own solipsism, he does feel a kinship with the stream,

> Reflecting every herb and drooping bud
> That overhung its quietness—'O stream!
> Whose source is inaccessibly profound,
> Whither do thy mysterious waters tend?
> Thou imagest my life.' (501–05)

The poet knows that the stream's source, and therefore the source of his own thoughts, are beyond his knowledge. However, he seems to believe that he can know the stream's end if he simply follows it far enough. But as the images of the poem show, going more deeply into one's own reflections leads to unawareness, solipsism, and death. By insisting that the stream is nothing more than an image of his own life, he fails to see through his own reflection and so hopes to see clearly by penetrating the reflection into the other side of the looking-glass, where death lies. His question can help him if he realizes that he must not pursue it to its ultimate conclusion, and must find a more imaginative way to pierce the reflection, as will happen in *Prometheus Unbound*.

Indeed this very wisdom may distinguish the older and wiser narrator from the mistaken poet. I have already shown a contrast in degree of reciprocity, but a deeper contrast illuminates the issue even more clearly. In the introduction to the poem, the narrator seeks to still the Wordsworthian "obstinate questionings" by finding out from ghosts "what we are" (26, 29). Instead of aggressively pursuing, creating, and therefore destroying, he acknowledges the need for a wise passiveness:

> ... moveless, as a long-forgotten lyre
> Suspended in the solitary dome
> Of some mysterious and deserted fane,
> I wait thy breath, Great Parent, that my strain
> May modulate with murmurs of the air
> (42–46)

We hope that he would not find himself in the same predicament as Coleridge if he yields to his grand metaphor of the Aeolian Harp. In any case, he understands that the human mind can go only so far and must then wait for reciprocal action from the outside world. The human mind can not create everything.

The final lines of "Mont Blanc," a poem which questions the relationship between the individual mind and the Power behind nature, are often interpreted as if they encouraged solipsism, but the lessons of *Alastor* should not be so easily lost.

> And what were thou, and earth, and stars, and sea,
> If to the human mind's imaginings
> Silence and solitude were vacancy? (142–44)

On its deepest level, of course, the question is unanswerable. If the human mind had no perceptions, as in the Shelleyan equation between internal ideas and external perceptions with which I opened this chapter, we would know even less than we know now and would have even less chance of answering the question. However, certain answers lead to dead-ends. If, for example, the Alastor poet were to answer, "Nothing, because all of nature exists only as a reflection of my own mind," then he hastens to his own sterile death, as we have seen.

Beyond the eyesight of man "the snows descend / Upon that Mountain; none beholds them there" (131–32), but they nevertheless exist beyond the perception of man. Man can see only the results of the process—"The fields, the lakes, the forests, and the streams" (84)—and not the power behind the process. Shelley is skeptical of any ability to perceive the Power, but his faith testifies that the Power exists: "the power is there" (127). As Pulos explains throughout *The Deep Truth,* Shelley is not undecided on the question, but is ruthlessly honest about the difficulties involved. Thus he ends the poem with a question which to some implies that all the power is in the perceiving mind.

However, the rest of the poem makes it clear that neither side of the relationship carries all the power. When the poet muses on his individual mind, he recognizes reciprocity, which was almost completely absent to the *Alastor* poet. More like the narrator of *Alastor,* the narrator of "Mont Blanc" sets up a dynamic back-and-forth between inner mind and outer universe:

> My own, my human mind, ... passively
> Now renders and receives fast influencings,
> Holding an unremitting interchange
> With the clear universe of things around. (37–40)

Two distinct aspects of the universe co-exist in "Mont Blanc":

> All things that move and breathe with toil and sound
> Are born and die; revolve, subside, and swell.
> Power dwells apart in its tranquillity,
> Remote, serene, and inaccessible. (94–97)

The third aspect of the poem, the perceiving mind, lives in the world of mutability. The inaccessible Power, symbolized by the snows on the top of Mont Blanc (line 131), exhibits itself, as Necessity, in the mutable world as a powerful river that fertilizes and as a powerful glacier that destroys. Like Shelley's West Wind, it is both destroyer and preserver. The "adverting mind" (line 100) learns from the world of mutability that the Power is indifferent to the welfare of man. However, properly understood,

> Thou hast a voice, great Mountain, to repeal
> Large codes of fraud and woe; not understood
> By all, but which the wise, and great, and good
> Interpret, or make felt, or deeply feel. (80–83)

The "voice" suggests the possibility of communion with a nature that possesses its own subjectivity. The minds of individual humans choose whether to listen to that voice, and then they choose to believe either that the Power creates or that it destroys. Because both are true, either belief can find verification in the outside world. In that sense, then, the mind does create its own reality. However, since some remote, indifferent Power does exist out there, either belief will sometimes find itself contradicted by the facts. The Power does not require man to be wise or great or good but if he is, then his world is much richer. If he believes that silence and solitude are vacancy, then he removes for himself the possibility of interacting and becomes a solipsistic *Alastor* poet. Without succumbing to the extreme of idealistic solipsism or the extreme of empirical pessimism, Shelley denies neither of them. Thus by fronting the complex truths of existence as honestly as he can, Shelley creates a poem of exquisite tension.

Earl Wasserman's reading of "Mont Blanc" brilliantly illuminates the relationship of subjective and objective and leads into the handling of the question in *Prometheus Unbound*:

> 'Silence and solitude,' those interstices between sensible ideas, are, Shelley wrote in *Mont Blanc*, a 'vacancy,' a hiatus creating that illusion of the discreteness and diversity of thoughts on which the individual mind subsists; and yet the purpose of that poem is to reveal, by means of an extraordinary apprehension of the imagination, that behind the illusory there are no such vacancies.[11]

Wasserman also explains that two opposing conceptions operate simultaneously in "Mont Blanc" as warp and woof of the poem.[12] On the one hand the river of things is said to be dark and glittering, but on the other hand such qualities are located in the mind. The reality which is perceived by the mind does have an objective existence, but our only perception of it is subjective. Therefore the idea is similar, but not quite the same as Wordsworth's "perceive and half-create" in "Tintern Abbey." So when Shelley writes that "nothing exists but as it is perceived," he does not mean that it ceases to exist if he does not perceive it, but that his only knowledge of its existence is through his perceptions.

The lessons of *Alastor* and "Mont Blanc" find their fulfillment in *Prometheus Unbound*. Although Shelley's poetry consistently shows that the perfect vision can not be attained by mortals, in *Prometheus Unbound* he enters the immortal realm in order to show the triumphant possibilities of his insights. In "Mont Blanc" it is clear that common experience does not let man know directly the Principle of Permanence,[13] although the poetic imagination can pierce through the clouds which veil the mountaintop and imagine the snowfall there. But an immortal who has lost full vision can regain it. Prometheus thus can stand for all mankind, and in fact Wasserman reads the character as the One Mind, of which all individual minds are portions. So when the reconciliation of mind and environment occurs in Prometheus, it occurs simultaneously in the whole universe.

Prometheus Unbound begins with what the sociologist Erik Erikson calls a "rock-bottom attitude," an attitude very similar to that of the poet at the end of *Alastor*.[14] There are of course many important differences between the two characters, not the least of which is the fact that the *Alastor* poet has not hated as Prometheus has, but they both seem to be at a dead end because of their refusal to see with anything but their single-minded faulty vision. The *Alastor* poet, however, has passed beyond any hope of recovery; Prometheus begins the play with a reversal of his hardened attitude toward Jupiter:

> Disdain! Ah no! I pity thee....
> ... I speak in grief,
> Not exultation, for I hate no more,
> As then ere misery made me wise. The curse
> Once breathed on thee I would recall.
> (I.53, 56–59)

Providing a kind of mirror image of the original curse, the curse is then repeated by the phantasm of Jupiter, which is a kind of mirror-image of Jupiter. Furthermore, throughout the play Jupiter is a cruel parody of Prometheus, to the extent that Jupiter is an anthropomorphic God made in Prometheus's image. Therefore, the mirror-image of the original curse is repeated by the phantasm of Jupiter, which is a mirror-image of Jupiter, who is a mirror-image of Prometheus. But the effect of these mirror-image reflections is the opposite of the effect in *Alastor*, for in that poem the reflections served to keep the poet from seeing the truth about himself and about the world. Here, on the contrary, the purpose of all these mirror-images is to make external that which is internal so that it may be understood and misperception corrected.

Barbara Gelpi, in her psychological explanation of *Prometheus Unbound*, sheds light on the nature of the recall of the curse. Although she emphasizes in her book that the center of critical interest in the interpretation of *Prometheus Unbound* has shifted from the recall of the curse at the beginning of the play to the appearance of Demogorgon in the center of the play, her evocation of the psychotherapy of Melanie Klein indirectly illuminates the dynamic creativity of Prometheus's recall. Referring to a little boy who was unable to talk because he was unable to express his sexual desire for his mother, Gelpi elaborates:

> for an adult subjectivity muted through neurotic conflict, the psychoanalytic process involves a renaming and calling forth of a subject who can call (i.e., conceive of and function within relationships) in a way not possible before. [Gelpi

> now quotes Klein:] "the interpretive gift is not constative (cognitive) but performative: the gift is not so much a gift of truth, of understanding or of meaning: it is, essentially, a gift of language.[15]

Without delving into the intricacies of Gelpi's entire argument about Shelley's and his society's attitudes toward mothers, I can point out that Prometheus's recall is, like Melanie Klein's patient's verbalization, a way for Prometheus to enter into dialogue, to escape from his solipsism, to come to terms with the intractable otherness of Jupiter, to change his ontological stance from one of pure, defensive, egotistical center to the back-and-forth movement of self and other. Only through a rigorous acceptance of the coincidence of opposites, of himself and Jupiter (remember that Earth soon thinks that he has capitulated because she does not imagine the breadth of his new understanding), does Prometheus allow the oneness of humanity to remain even while he adds at the same time the otherness of alterity.

Before he recalls the curse, Prometheus accuses Jupiter of "self-torturing solitude" (I.295). That stance, which was also the stance of the *Alastor* poet, is the one from which Prometheus is freeing himself. Just as in *Alastor*, such a stance must lead to its inevitable conclusion, revelation:

> An awful image of calm power
> Though now thou sittest, let the hour
> Come, when thou must appear to be
> That which thou art internally. (I.296–99)

A similar kind of inevitable reversal appears in an image drawn from nature. When Jupiter first appears, he brags that man's soul, although not yet conquered, must inevitably be conquered, because the curses of Jupiter will pile up like snow:

> ... my curses through the pendulous air
> Like snow on herbless peaks, fall flake by flake
> And cling to it [the as yet unsubdued soul of man]
> (III.i.11–13)

Jupiter, however, misunderstands the real meaning of the power of this natural image. We as readers know that Jupiter's words ironically echo some of Asia's words just two scenes earlier:

> *Asia*: ... Hark! the rushing snow!
> The sun-awakened avalanche! whose mass,

> Thrice sifted by the storm, had gathered there
> Flake after flake, in Heaven-defying minds
> As thought by thought is piled, till some great truth
> Is loosened, and the nations echo round
> Shaken to their roots: as do the mountains now.
> (II.iii.36–42)

Jupiter forgets that cycles contain declines as well as ascents; he forgets that his power is not eternal, but is only part of the natural ups and downs of nature. Thus Shelley employs images from nature to reassure his readers of the inevitable downfall of tyrants, as he does in his sonnet "Ozymandias." As we shall see in Blake's myth, the cycles of nature do not offer such hope but only the promise of inevitable repetition unless escaped through the power of the imagination. But in Shelley the combination of the inevitability of the natural cycle and the inevitability of inner truth becoming visible provides the undercurrent of hope that drives the whole poem.

The process of making what is invisible and internal become visible and external so that it may stand or fall on its own weight and not with the protection of veils, lies, or hypocrisies, is absolutely fundamental to Shelley's conception of *Prometheus Unbound*. In the preface he claims that

> The imagery which I have employed will be found, in many instances, to have been drawn from the operations of the human mind, or from those external actions by which they are expressed. (p. 133)

Once again Shelley treads the fine line between external and internal, turning one into the other. Thus, when the Furies besiege Prometheus, he fears that such a process may be reversed and the external become internal:

> Methinks I grow like what I contemplate,
> And laugh and stare in loathsome sympathy.
> (I.450–51)

As Blake so often intones, we can become what we behold; if Prometheus has stopped beholding the tyranny of Jupiter, he may become caught in another staring contest, in which he sees his own horrifying reflection. The spiritual torture that the Furies inflict on him is also a physical torture that invades his body.

Thus internal thought finds its objective correlative in images of the physical body:

> Thou think'st we will live through thee one by one,
> Like animal life, and though we can obscure not
> The soul which burns within, that we will dwell
> Beside it, like a vain loud multitude
> Vexing the self-content of wisest men:
> That we will be dread thought beneath thy brain,
> And foul desire round thine astonished heart,
> And blood within thy labyrinthine veins,
> Crawling like agony?
> *Prometheus*: Why, ye are thus now;
> Yet am I king over myself, and rule
> The torturing and conflicting throngs within.
> (I.483–94)

Thus it seems that Prometheus's corrective for the torture which he has undergone is not to deny or ignore it, but to incorporate it into his total being just as a king reconciles disparate elements, and even rebellious ones, within his kingdom. The mind, represented by Prometheus, finds itself in long-lasting combat with something outside itself, represented by Jupiter. However, once the mind recognizes that the opposition is created by its own perception, it can use its imagination to transcend the simple opposition. By recognizing its own part in maintaining the battle, and bringing all possible truths to light, the mind can choose to "no longer abide by the standards of antagonism,"[16] as Stuart Curran puts it. Putting the question in a Miltonic perspective, he explains that Shelley saw Milton using a similar model at the center of *Paradise Lost*. Because both Satan and God contain admirable and despicable characteristics, Shelley saw a conceptual framework that transcended narrow moral questions, just as he was trying to do.

The Shelleyan attempt at synthesis does not conceal the difficulties. Good intentions in the mind do not necessarily succeed in the world, as the Furies remind Prometheus in an attempt to discourage him:

> The good want power, but to weep barren tears.
> The powerful goodness want: worse need for them.
> The wise want love; and those who love want wisdom;

And all best things are thus confused to ill.
Many are strong and rich, and would be just,
But live among their suffering fellow-men
As if none felt; they know not what they do.
(I.625–31)

Could there be a more gloomy and discouraging picture of human existence? Not even Yeats's similar lines in "The Second Coming" imply such pessimism. How can Shelleyan optimism possibly overcome such self-opposition? The good that men do is sorely and consistently opposed, and almost negated, by the bad.

Like Blake, but in a less radical way, Shelley sets up a dreadful, impossible situation, but then shows that inextricably contained within that dreadful situation is the means of solution. By the logic of the coincidence of opposites, if evil always infects good, if cruelty and pain are inextricably bound up with kindness and comfort, if the best situation contains also the worst, then likewise the worst also contains the best. Just after the pessimistic passage quoted above, Shelley gives us the most radiant images of hope, which will begin to usher in the new age of Act IV. A chorus of spirits arrives to announce, among other events, that a sailor "gave an enemy / His plank, then plunged aside to die" (IV.i.721–22). This extreme gesture of kindness to the opponent highlights Prometheus's moral imagination in ending the antagonism between himself and Jupiter without yielding. As in the story of the noble fellow in *Frankenstein* (see discussion in previous chapter) a possible solution to the stand-off is to deny the interests of the self to the extreme of the self-annihilation of putting another self before yours. Of course Prometheus's (and by implication Shelley's) imaginative answer has managed to maintain the integrity of the self even while denying the stand-off. But the sailor's gesture underlines the element of self-sacrifice necessary to the solution. In a sense the sailor becomes the other—in the figurative sense that his act of generosity springs from empathy and in the literal sense that he dies instead of his enemy. We shall see Blake developing this self-sacrifice—"every kindness to another is a little Death"—to an even greater extreme.

The next step in Prometheus's escape from the stand-off of apparently mutually exclusive oppositions, after he has recalled his curse against Jupiter, is to join with an otherness that has been missing, his female counterpart, Asia. His opposition to Jupiter has been similar to the stand-off between Los and Spectre in Blake's *Jerusalem*. The Spectre, like Jupiter, an obvious hindrance, opponent, and even torturer, must become a forgiven part of the self. Asia, like Jerusalem to Albion, must also become a fundamental part, but her role rather than that of an opponent, is at first that of an absent person. After confronting his reflecting trap with Jupiter, Prometheus is ready to join Asia, who has been waiting for him in a once desolate vale "now invested with fair flowers and herbs" because of "her transforming presence, which would fade / If it were mingled not with thine" (I.829, 832–33). As in the Blakean myth, the male must be united with the female as part of the total re-creation of unity. Asia is not a mere reflection of Prometheus's mind, as the female in *Alastor* was of the poet's mind, but a genuine other who can enter into a synthesis with him.

When Asia enters the play, she immediately demonstrates her ability to pierce through surfaces and reflections to see essentials. She does not get caught in mirror surfaces and their solipsistic implications, as do Eve in *Paradise Lost*, the creature in *Frankenstein*, and the poet in *Alastor*. She does not get caught in the kind of trap that pitted Prometheus against Jupiter for eons. Her sister Panthea, coming to tell her of a dream about Prometheus, assumes a limited, mirror-image perspective as the only possible goal of vision when Asia tries to see beyond the words she can hear:

> *Asia.* Thou speakest, but thy words
> Are as the air. I feel them not....
> [ellipsis in original] Oh lift
> Thine eyes, that I may read his written soul!
>
> *Panthea.* I lift them, though they droop beneath the load
> Of that they would express—what canst thou see
> But thine own fairest shadow imaged there?
>
> *Asia.* Thine eyes are like the deep, blue, boundless Heaven
> Contracted to two circles underneath
> Their long, fine lashes—dark, far, measureless,—
> Orb within orb, and line through line inwoven.—
> ...

> There is a change: Beyond their inmost depth
> I see a shade, a shape: 'tis He. (II.i.109–20)

This fantastic image sharply sets the contrast between two modes of vision: one mode looks at another's eyes and sees only a reflection of the self as the *Alastor* poet would and as Prometheus was trapped into doing before the play began; the other mode sees through the reflection into a deeper truth, as the poet of "Mont Blanc" attempted to do. Of course a reflection in another's eyes can always be seen if the perceiver assumes a certain focus. However, more difficult and more important is the image when one sees through the reflection. Such perception enables a genuine interpenetration instead of a mere bouncing off of surfaces. Asia thus perceives the other with whom she will establish a reciprocating unity, not a solipsistic one. Asia's ability to perceive here provides an illuminating contrast with the *Alastor* poet who looks into a well and sees not only his own reflection, but less than that, the reflection of his own eyes. Reflection, whether that of Eve, or of Frankenstein's creature, or of Panthea, allows a vision only of the solipsistic self; penetration enables a blending, an interaction, a coincidence of opposites.

However, Asia is about to be led to even deeper truth, a truth beyond the imagery of sight and reflection, as she descends into the cave of Demogorgon and asks him difficult and sometimes unanswerable questions.

> *Demogorgon.* Ask what thou wouldst know.
> *Asia.* What canst thou tell?
> *Demogorgon.* All things thou dar'st demand.
> (II.iv.8–9)

But when she presses her questions to the point of asking about

> Abandoned hope, and love that turns to hate;
> And self-contempt, bitterer to drink than blood,
> (II.iv.24–25)

mysteries similar to those gloated over by the torturing Furies, the answer "God," which has been given to several questions, changes into "He reigns." In a penetrating psychological and spiritual history of man, which emphasizes Prometheus's role as creator and synthesizer, Asia leads up to the question of who is above Jupiter. Demogorgon's answer is the heart of this play and perhaps of Shelley's thought:

> the deep truth is imageless.
> (II.iv.116)

Although the rest of the play is one of the very few successful visions of apocalypse in literature, still the truth lies more deeply. After all the attempts in the images created by language, still the essential mystery remains. Shelley may be attaining here a stance similar to one he expresses in his essay "On Life":

> How vain is it to think that words can penetrate the mystery of our being. Rightly used they may make evident our ignorance to ourselves, and this is much. (pp. 475–76)

Sounding strangely like Nicolas Cusanus in his learned ignorance, Shelley there seems content to rest without trying to penetrate further. In *Prometheus Unbound,* however, the urging to penetrate continues. Asia has one more question for Demogorgon:

> One more demand; and do thou answer me
> As mine own soul would answer, did it know
> That which I ask. Prometheus shall arise
> Henceforth the sun of this rejoicing world:
> When shall the destined hour arrive?
>
> *Demogorgon.* Behold!
> (II.iv.124–28)

And finally, of course, this must be the answer to the deepest questions. Asia's question can be translated: When will the ideal vision in the mind and the reality of the external world coincide? And Demogorgon's answer can be interpreted to mean: It will happen when you look and see it happening. His answer is not "Now," but rather "Look and see." For the answer must come from Asia's own soul, as she knows, just as Prometheus's answer had to arise from his own soul.

Until the mind is able to transform its habitual reflexive vision into penetrating imaginative vision, it will remain trapped in old patterns of frustration and despair. But the mind does not create external reality, as the *Alastor* poet tried to do, and as Coleridge tried to do. Rather, the mind voluntarily and imaginatively enters into dialogue with what appear to be recalcitrant forces outside the self. Thereby the perceiving mind and the external world are transformed.

> According to a traditional interpretation, when Saint John says 'Come and see,' he refers at once to the tenor of his prophecy—a revolution in vision—and its vehicle, vision itself. The structures and techniques of prophecy all serve to make its form match its content. On one level, then, a prophecy *means* the intellectual efforts that it *exacts*.[17]

The rest of *Prometheus Unbound* abounds in images of glory and joy as the new age dawns. On its simplest level a new kind of vision simply disregards the old tyrannies. Sceptres, swords, palaces, and other symbols "Stand, not o'erthrown, but unregarded now" (III.iii.179). On a more problematic level, eternal vigilance is the price of the newfound glorious vision:

> To suffer woes which Hope thinks infinite;
> To forgive wrongs darker than death or night;
> To defy Power, which seems omnipotent;
> To love, and bear; to hope till Hope creates
> From its own wreck the thing it contemplates;
> Neither to change, nor falter, nor repent;
> This, like thy glory, Titan, is to be
> Good, great and joyous, beautiful and free;
> This is alone Life, Joy, Empire, and Victory.
> (IV.570–78)

The triumph at the end of *Prometheus Unbound* is thus not a once-and-for-all victory that ends on Easy Street. It is a hard-fought, continuous struggle that fully acknowledges the exterior and the interior forces that would destroy it. If inevitable cycles could bring Jupiter down, then inevitable cycles can bring him or something worse up again, as in Yeats's "The Second Coming." But Demogorgon's benediction provides a dynamic, continual re-creation of existence, rather than a Jupiterean triumphant bragging smugness. Even if tyranny returns, and it always will return, the creative imagination does not have to suffer defeat.

In *The Triumph of Life* images of infinite woes and dark wrongs predominate, but images of hope, even though muted, dissolved, and washed away, do not completely disappear. The unmistakable pessimism of the poem leads Carl Grabo to declare that it shows complete disillusionment with earthly life, like that of an old man about to die.[18] Donald Reiman, on the other hand, sees that positive alternatives balance the negative images, and he even goes so far as to say that in this poem Shelley "renounced his pragmatic compromise with amoral Necessity" demonstrated by his compromise with terrestrial limitations in *Ode to the West Wind, The Mask of Anarchy*, and other revolutionary poems, "and turned once again to the elucidation of the Ideal," as in *Prometheus Unbound*.[19]

More compatible with my approach is the idea of Michael O'Neill:

> *The Triumph of Life* neither rejects nor re-affirms the poet's Promethean faith in the human spirit. Instead, it tests it.[20]

I believe that a further development of that approach to *The Triumph of Life* can begin with a letter that Shelley wrote to Horace Smith on 29 June 1822, just nine days before his death, probably while he was in the throes of composing *The Triumph of Life*:

> Let us see the truth, whatever that may be. The destiny of man can scarcely be so degraded that he was born only to die—and if such should be the case, delusions, especially the gross and preposterous ones of the existing religion, can scarcely be supposed to exalt it. If every man said what he thought, it could not subsist a day.[21]

Shelley's determination to see the truth, certainly a major impetus behind *Alastor*, "Mont Blanc," and *Prometheus Unbound*, finds its most dramatic and starkest presentation in *The Triumph of Life*. The poem opens with what appears to be a very favorable image of the sun in the morning, but when all of nature begins to worship it, the language sounds very much like that used by Shelley to condemn tyranny. Further to undercut the sun image, Shelley has his poet-narrator separate himself quietly but unmistakably from the worship of the sun. This kind of balance, or perhaps ambiguity, continues as the narrator places himself precisely on the borderlines of time and space:

> before me fled
> The night; behind me rose the day; the Deep
> Was at my feet, & Heaven above my head. (26–28)

Like the Dante whom he was coming to imitate more and more, the poet begins in the middle of his way on earth: "Nel mezzo del cammin di nostra vita." In this condition, balanced between the most fundamental opposites of earthly existence—night and day, ocean and sky—the poet sees his vision. Following a miserable multitude comes a chariot which

> Like the young moon
> ...
> Doth, as a herald of its coming, bear
> The ghost of her dead Mother, whose dim form
> Bends in dark ether from her infant's chair. (79, 83–85)

Besides echoing Shelley's own moon imagery in *Alastor*, this image is reminiscent of a similar image at the opening of Coleridge's *Dejection Ode*, which, as M.H. Abrams expresses it, concerns "the loss of the mind's reciprocating power"[22] with nature. Coleridge begins by quoting an old ballad, and then describing his vision of the moon:

> Late, late yestreen I saw the new Moon,
> With the old Moon in her arms.
> ...
> Ballad of Sir Patrick Spence
>
> ...
>
> ...
> For lo! the New-moon winter-bright!
> And overspread with phantom light,
> ...
> I see the old Moon in her lap, foretelling
> The coming-on of rain and squally blast.

And indeed *The Triumph of Life* calls into question any favorable reciprocal relationship between mind and nature by showing Jean-Jacques Rousseau complaining bitterly about the inadequacy of the exterior world's diet for his idealistic spirit. In Shelley's caricature of Rousseau, the founder of Romanticism simply can not get enough spiritual food to eat:

> And if the spark with which Heaven lit my spirit
> Earth had with purer nutriment supplied
>
> Corruption would not now thus much inherit
> Of what was once Rousseau. (201–04)

When first met earlier, Rousseau appears as an old stump, as if regressed into Blakean single vision[23]:

> ... what I thought was an old root which grew
> To strange distortion out of the hill side
> ... the grass which methought hung so wide
> And white, was but his thin discoloured hair,
> And ... the holes it vainly sought to hide
> Were or had been eyes. (182–88)

Harold Bloom sees Rousseau in just these Blakean terms, as a human who has been diminished by becoming part of nature:

> Rousseau has attained completely to the state of nature; grotesquely, he has become a part of what Blake calls the Vegetative universe ... rather than a natural object having been humanized, a human being has been naturalized.[24]

E. Douka Kabitoglou, on the other hand, disagrees with that derogatory reading and thinks that Rousseau is an "appropriate mediational figure to initiate the Poet into the mysteries and meaning of the universe—and life."[25] In images similar to those of the *Alastor* poet hastening to his death, Rousseau is shown defeated by Life. In fact the chariot of Life seems to frustrate or destroy all except those who die young or refuse earthly glory and wealth (128–33).

The poet's sympathy for the miserable multitude leads him to notice that

> none seemed to know
> Whither he went, or whence he came, or why
> He made one of the multitude. (47–49)

The questions of whither, whence, and why, and variations of them, form the center of the poem, as they are repeated several times.[26] Paul de Man points out that these series of almost repeated questions are "a more subject-oriented, second-person mode" of "the questions that articulate one of the text's main structures: '... "And what is this? / Whose shape is that within the car? and why—"' (ll. 177–78)."[27]

As soon as the poet has a chance, he asks his vision of Rousseau about the relationship of the individual to the parade of Life. He wants to know the reasons for such a discouraging "flow / Of people" but most of all, he wants to know Rousseau's origin and goal:

> "Whence camest thou & whither goest thou?
> How did thy course begin," I said, "& why?" (296–97)

Rousseau responds that he does not know where he is going or why, but suggests, probably sarcastically, that the poet might want to join the procession and find out for himself (305–08). However, judging from his earlier advice to forbear the dance (188–89) and his peremptory "now listen" (308), it seems that Rousseau believes that the poet will be better off if he hears the story instead of living it. In fact, if the poem contains any hope at all, surely part of that hope must be that there is some imaginative alternative to the grim procession of Life.

As Rousseau relates his own story, he reveals that he has asked similar questions himself. After awaking in a Wordsworthian world from which some undefined glory seems to have faded, he perceives a "shape all light" (352) created from the interaction of sunlight on well-water, and therefore probably a product of his own imagination. Rousseau inquires of this vision within a vision:

> If, as it doth seem,
> Thou comest from the realm without a name,
>
> Into this valley of perpetual dream,
> Shew whence I came, and where I am, and why. (395–98)

Rousseau is a step ahead of the *Alastor* poet, who, upon awaking from his imagined love-making with the woman of his fanciful creation, was not able to ask questions as penetrating as these. The response to Rousseau's questions is "Arise and quench thy thirst" (400). This answer is similar to the one that we saw Demogorgon give to Asia: "Behold." But Rousseau is being urged to do what he, after this experience, advised the narrator of *The Triumph of Life* not to do. As soon as he follows the injunction to drink, Rousseau's mind finds its perceptions altered:

> And suddenly my brain became as sand
>
> Where the first wave had more than half erased
> The track of deer on desert Labrador,
> Whilst the fierce wolf from which they fled amazed

Leaves his stamp visibly upon the shore
Until the second bursts—so on my sight
Burst a new Vision never seen before. (405–11)[28]

This new vision is the chariot of Life. In crude allegory, the facts of empirical existence, represented by the wolf tracks, have replaced the idealistic vision, represented by the deer tracks. However, the deer tracks have been only partially erased, and so some vestiges of hope remain. Further hope is contained in the assurance that a future wave will erase the wolf tracks. As in "Mont Blanc" and *Prometheus Unbound,* natural processes not only bring about the reign of tyranny, but they promise its eventual downfall. In several other images, some of which are reminiscent of those in *Alastor* referring to the poet's ideal vision, Rousseau's former vision remains a yearned-for ideal as he joins the procession around the chariot of Life.

In both examples of a series of questions, the questioner does not receive the answers that he demands. This fact suggests two possibilities: either the questions have no answers, because the deep truth is imageless as in *Prometheus Unbound,* or the questions are not the ones that should be asked. Both of these possibilities illuminate reasonable meanings of the poem. The latter possibility arises when Rousseau distinguishes himself from the "spoilers spoiled" (235):

For in the battle Life & they did wage
She remained conqueror—I was overcome
By my own heart alone. (239–41)

These lines suggest that the important battle is waged inside the individual. Instead of worrying about origins and final goals, instead of resisting and complaining about the unfairness of the world, perhaps it is better to understand the self and meet the world with firm mind, much as Prometheus does when he rules the kingdom within himself. That message is even more explicit a little earlier in the poem:

The Wise,
The great, the unforgotten … their lore
Taught them not this—to know themselves; their might
Could not repress the mutiny within,
And for the morn of truth they feigned, deep night
Caught them ere evening. (208–15)

Unlike Prometheus, they can not control the political rebellion within the self, in order to reconcile discordant elements, to make opposites coincide in one body. Added to the image of the kingdom of the self is an important image of night and day. The obvious, or daylight side of their personalities was a false one, because it did not include the recalcitrant elements within. The darker, or night side conquered them because "deep night" contained a deeper truth than did the feigned "truth of morn."

But an even darker possibility awaits: perhaps the "mutiny within" has power to win because the questions have no answers. The many shifts of perspectives, the visions within visions, the fading of one thing into another, the images of dissolution and destruction, all contribute to a feeling of uncertainty, in which no sure answers exist. This destruction of certainties was an important part of Romantic poetry in general and of Shelley's poetry in particular. In *Alastor* we see the triumph of a reflection that destroys because it does not include the creative imagination. In "Mont Blanc" and *Prometheus Unbound* the imagination struggles toward and succeeds in accomplishing a synthesis of subject and object. In *The Triumph of Life* the necessary destruction and dissolution of old forms occurs. But because Shelley died without finishing the poem, it ends with the question, "What is Life?" Perhaps Shelley would have shown that the question is the wrong one to ask; perhaps he would have shown that it has no answer.

Many critics think that *The Triumph of Life* would have continued optimistically, but agree with Edward Bostetter that the fragment as left reveals disenchantment, and unlike Shelley's other poetry, contains "no hint of optimistic resolution."[29] Judith Chernaik sees no possibility of optimism, and explains that Shelley's last lyrics confirm the darkening of spirit that is suggested so strongly by *The Triumph of Life.*

> One senses an increasing pessimism on Shelley's part, the gradual withdrawal of the poet's bright Elysium until it appears to be no more than an image of what the soul in vain desires, counter to all reason, all evidence, desires.[30]

John Hodgson sees "the apparently linear triumph of Life" absorb "the cycles of human life and death and time."[31] P.M.S. Dawson points out that the poem is "remarkably dense with allusion ... to a number of natural cycles,"[32] the largest of which is the comparison of human life to the seasons. This analogy suggests that the optimistic ending of "Ode to the West Wind" may be relevant here. But Dawson calls applying natural cycles to human life an error as he concludes:

> The recognition to which the figures in *The Triumph of Life* do not attain, at least within the poem as we have it, but which the poem offers to its readers, is that the history of which human beings seem to be the helpless victims is their own creation—our own creation.[33]

In a distinct minority, Kabitoglou asserts that the poem is not a fragment and that "the interrogative form that nearly closes it is the only manner in which a poem teeming with ambiguities and 'overwhelming' questionings could end"[34]:

> The poem's final asking of what is the meaning of life is not only a thrust into the void but a definite progress, a minor(?) [sic] triumph over the ineffable, in that it substitutes the earlier question asked—'And what is this?'—by an interrogative type that partly contains its own answer, in the self-referentiality of the final question form. The answer is not something 'out there'; the answer to the one obstinate question of what it is all about, is necessarily the asking of it, the aporia—the living and experiencing the perplexity.[35]

Faithful to Shelley's image of cycles, Rubin suggests that the poem shows us a kind of lower limit, which it would have worked its way out of had Shelley lived. Referring to line 469, Rubin writes,

> the 'wonder' which Rousseau is about to describe is the last sight that is revealed before the chariot begins its upward climb out of the mysterious dell. There is some suggestion here that the chariot must have descended to the nadir of its journey, so that we are given to feel that past this last, impending, horrific sight, it can no farther fall and that the next move must be an ascension of some kind.[36]

In any case, Shelley does not give us, finally, a coincidence of opposites like the one attained in *Prometheus Unbound*, but he does give us a struggle between inner and outer that asks many of the same questions that Blake asks. And Blake will give us answers in the form of the coincidence of opposites.

Notes

1. All quotations of Percy Bysshe Shelley are taken from Donald Reiman and Sharon Powers, eds., *Shelley's Poetry and Prose* (NY: Norton, 1977).

2. "On Life", pp. 477–78.

3. C.E. Pulos, *The Deep Truth: A Study of Shelley's Scepticism*. University of Nebraska Press, 1954.

4. Merle R. Rubin, "Shelley's Skepticism: A Detachment Beyond Despair," *Philological Quarterly*, 59 (1980), p. 356.

5. "On Life," p. 476.

6. M.H. Abrams, *The Mirror and the Lamp* (London: Oxford UP, 1953), pp. 168–77.

7. Abrams, p. 53.

8. Rubin, pp. 354, 356.

9. *Byron's Poetry*, ed. Frank McConnell (Norton, 1978), p. 139.

10. *William Wordsworth*, ed. Stephen Gill (Oxford UP, 1984), p. 297.

11. Wasserman, *Subtler Language*, p. 198.

12. Wasserman, p. 200.

13. Wasserman, p. 226.

14. W.H. Hildebrand, "Shelley's Early Vision Poems," *Studies in Romanticism*, 8 (1969), p. 211.

15. Barbara Charlesworth Gelpi, *Shelley's Goddess: Maternity, Language, Subjectivity* (NY: Oxford UP, 1992), p. 213.

16. Stuart Curran, "The Siege of Hateful Contraries," *Milton and the Line of Vision*, ed. J.A. Wittreich (Madison: Univ. of Wisconsin Press, n.d.), p. 218.

17. Hoagwood, p. 11.

18. Carl Grabo, *The Magic Plant: The Growth of Shelley's Thought* (Chapel Hill: Univ. of North Carolina Press, 1936), p. 421.

19. Donald Reiman, *Shelley's "The Triumph of Life": A Critical Study* (Urbana: Univ. of Illinois Press, 1965), p. 84.

20. Michael O'Neill, "Shelley's *The Triumph of Life,*" in Watson, *Infinite*, p. 161.

21. Holmes, *Shelley*, p. 727.

22. Abrams, *Mirror*, p. 67.

23. See discussion of single and double vision in Chapter 1.

24. Bloom, *Shelley's Mythmaking*, p. 255.

25. *Plato and the English Romantics: Dialogoi* (NY: Routledge, 1990), p. 110.

26. P.M.S. Dawson emphasizes many repetitions in the poem, including the feeling of *déja vu* in lines 33–37. Perhaps the poet is "literally repeating the experience of Rousseau," making the two characters into one in the same sense that Victor Frankenstein and his creature are one. ("The Mask of Darkness," in Behrendt, *History*, p. 237).

27. "Shelley Disfigured," *Deconstruction and Criticism*, ed. Harold Bloom (NY: Seabury Press, 1979), p. 39.

28. O'Neill provides a Blakean epigram: "consciousness altering, alters all." (p. 161)

29. Bostetter, p. 182.

30. Chernaik, p. 178.

31. Hodgson, p. 91.

32. Dawson, p. 239.

33. Dawson, p. 243.

34. Kabitoglou, p. 122.

35. Kabitoglou, p. 123.

36. Rubin, p. 359.

Chapter 6

Ways of Escape: Blake's "The Mental Traveller"

William Blake's "The Mental Traveller" serves as a good introduction to the coincidence of opposites in Blake because it shows the absolute failure of opposites to interpenetrate. The opposing male and female in the poem frustrate, dominate, and torture each other at every opportunity. Moving contrapuntally with the negations between fixed male and female principles, in a more mobile opposition, youth and old age mutually torture and attempt to destroy each other. The images of circles and of opposing centrifugal and centripetal forces that were crucial to Cusanus, Pope, Coleridge, Percy Shelley, and Mary Shelley, here take over completely. Circles generated by the insistent opposition between male and female threaten to end existence by their claustrophobic escalations of tension. Before Blake finds solutions to the problems of opposites in his later poems, he first paints them, in all their detail and agony, in "The Mental Traveller."

The oppositions in "The Mental Traveller" create a poem that well serves as a cautionary tale in one's methods of interpretation. At one extreme, some interpretations emphasize the horror of the cycle of male-female domination; at the other extreme, some interpretations emphasize the eternal aspects of the cycle of spirit and nature. I prefer the point of view of Michael Cooke, who emphasizes the poem's curious doubleness:

> An atmosphere of outrage at the entire scene pervades the poem, but there is also an uncontrollable fascination that the speaker freely imparts.[1]

All major forms of interpretation emphasize or imply the inevitability of the cycle, while only a few critics suggest any hope of escape. I maintain that the poem opens possibilities of escape in almost every moment. An even more optimistic view of the poem comes from Rachel Billigheimer:

> Blake employs imagination in order to escape the wheel of time.... In the historical-mythic account of "The Mental Traveller" Blake symbolically describes how freedom is born from suffering that is turned into triumph.[2]

I think that Billigheimer goes too far in reading this poem as exhibiting the triumph of the imagination. But I do agree that the excruciating cycles of this poem point toward the freedom of the imagination that blossoms in later poems.

Quite clearly the characters in the poem, or perhaps more exactly the principles of action in the poem, see opposites only as mutually exclusive. All attempts at interpenetration of male and female result in exclusion, torture, or destruction. Not accepting any co-existence of opposites, not accepting any mutually productive dynamic, not accepting any acts of inclusion, the male and female can try only to exclude or destroy or overpower or dominate. Imagination at any point might find freedom from the cyclical trap; instead the trap grows deeper and more horrible. When the poem ends—"And all is done as I have told"—the anticipated cycle promises to be even more brutal than the one in which we have just been spun.

Throughout the cycle, however, opportunities for escape abound. Most commentators recognize the promise of liberation in the birth at the beginning of the poem. It seems that the torture of the babe by the old woman might be avoidable. Even though reading the beginning of the poem in light of the end with its promise of repetition tends to dampen such a hope, the entire poem presents itself in a series of glimpsed, missed opportunities. At every moment the perspective of the imagination is possible, as it is in Blake's *Milton*:

> There is a Moment in each Day that Satan cannot find
> Nor can his Watch Fiends find it, but the Industrious find
> This Moment & it multiply. & when it once is found
> It renovates every Moment of the Day if rightly placed.
> (35:42–45, E136)

Putting this possibility in the context of Blake's entire œuvre, without specific reference to "The Mental Traveller," Thomas Altizer suggests a most radical hope based on acceptance of despair:

> The movement from Fall to Apocalypse is a dialectical movement through an 'Eternal Circle' demanding a full participation in every turn of the wheel.... Apart from the joy and horror of our fallen history, there could be neither a real nor a dialectical movement culminating in the Apocalypse. Therefore, every moment not only opens into Eden, but also the actual reality of Eden is inseparable from a fallen time and space. [3]

Whether or not we accept such an extreme insight, clearly Blake calls us in this poem to some such breaking of the normal limits of perception. Altizer's emphasis on the inseparability of fallen vision and Edenic vision, with its acceptance of the horrors of the cycle, can help us to reassess our disapproval of torture in the poem. Nevertheless, we must guard against a complacency that can be caused by Altizer's death-of-God theology. Simple acceptance of the horrors is not enough; it must be combined with an equal and opposite desire for escape. Blakean vision simultaneously accepts and rejects the everyday world of fallen vision.

From the very beginning of Blake's illuminated works, in his tractates, he distinguishes between poetic and empirical modes of vision. Probably the most useful general way to interpret "The Mental Traveller" is to apply Blake's conclusion to the tractate "There is No Natural Religion":

> If it were not for the Poetic or Prophetic character the Philosophic & Experimental would be seen at the ratio of all things, & stand still unable to do other than repeat the same dull round over again. (E1)

Ordinary vision, single vision, everyday common sense, called "Philosophic & Experimental" by Blake, can live only a numbing cycle. That cycle is perfectly represented in "The Mental Traveller," in which the characters and many interpreters can see no way to escape. The poet/prophet Blake, through his narrator who has travelled through these cycles, gives us the most horrifying picture of the philosophic/experimental limitations on existence. Mere empiricism, coupled with rationalism, contains no spark with which to light the psychological and spiritual darkness that envelops the cycles of "The Mental Traveller."

Like Blake's character Urizen when at his worst, the characters in "The Mental Traveller" fear any change brought on by new life:

> Urizen can understand recurrence well enough, but the presence within time and space of life, of a power which grows and alters its form, inspires in him a feeling of insecurity.[4]

Caught in unquestioned ideologies, caught in the traps of accepting the natural world as a standard, such limited vision can only accept the revenge that makes the world go round. But each action, each state or condition in the poem, presents a typical Blakean hope concealed within the apparent hopelessness. Each crux in the poem can be interpreted as a hopeful possible way of escape. True, each time they have the opportunity, the characters choose not to see Eternity, because each time the male or the female attains a potential for escape, the other negates it. However, this cycle of despair, of failed hopes, contains the germ of regeneration. This poem does not show such regeneration, as do *The Four Zoas*, *Milton*, and *Jerusalem*, but it does show the problem in stark opposites that never find coincidence. It does define opposites that need interpenetration, but that always seek it only perversely.

Blake's concept of States and Individuals also gives us a useful general framework in which to place "The Mental Traveller":

> Distinguish therefore States from Individuals in those States.
> States Change: but Individual Identities never change nor cease.
>
> (*Milton* 32:22–23, E132)

> The Spiritual States of the Soul are all Eternal
> Distinguish between the man, & his present State.
>
> (*Jerusalem* 52, E198)

> So Men pass on: but States remain permanent for ever.
>
> (*Jerusalem* 73:45, E229)

Although these passages may seem to contradict each other in their emphasis on whether states or individuals change, they all emphasize the dynamic nature of human identity. "Every harlot was a virgin once." The problem in "The Mental Traveller" can be defined as an insistence by the characters that everyone stay in the same condition. The actions of the characters try to fix the other or the self in a permanent condition. Ironically, as the cycle proceeds, such fixity is impossible anyway; the obvious lesson is never learned.

As the poem proceeds, the characters undergo dynamic changes of condition, but they always try to hold on to the state in which they find themselves, to forestall any further changes. The turning cycle drives them outward with centrifugal force, forcing change upon them; their selfishness of limited vision counters with centripetal force, desperately and uselessly grasping and clinging. At almost every turn they grasp, bind, and otherwise try to fix the state of the other individual and therefore of themselves. This grasping ironically produces an effect not wished for: instead of freezing the cycle, such grasping spins it faster.

The poem begins by re-casting the traditional word of human universality—"Men"—into a word of sexual division—"Men & Women." These bland, acceptable terms, beginning in line one with hegemonical unity, bifurcating in line two into a cheery "vive la différence," suddenly turn threatening and hostile as the narrator intrudes "dreadful" and "cold" into our world of traditional male dominance and sexual flirtation:

> I traveld thro' a Land of Men
> A Land of Men & Women too
> And heard & saw such dreadful things
> As cold Earth wanderers never knew.

After a first line which sounds like a routine travel memoir—"I traveld thro' a Land of Men"—the second line repeats half of the first line and adds an important split: "A Land of Men & Women too." In ordinary usage the first line about men would include women. By emphasizing the two sexes Blake underlines a drastic difference, an irreconcilable opposition that prepares us for the horrors of sexual separation that the rest of the poem catalogues. He may also be including a little joke like the one that

Hamlet tells at the end of his "What a piece of work is man!" speech:

> man delights not me; no, nor woman neither, though by your smiling you seem to say so.[5]

The traditional obliteration of woman in the universal term "man" not only denigrates women by excluding them, but it also highlights the licentious flirtation that Hamlet and Blake imply, the flirtation of the battle between the sexes that spirals into sadism in "The Mental Traveller." Women are the playthings of men in Hamlet's joke, the easy answer to a difficult question: if life is getting you down, go get yourself a woman. Women are the playthings of men in Blake's poem as well, but in this vicious equal-opportunity cycle, men are also the playthings of women. Without the difference in the sexes, the world could not go round; without the mutual exploitation, the cycles of "The Mental Traveller" could not continue in the same mutually destructive fashion. Although Blake does underline the categories of male and female in a peculiarly twentieth-century way, his point is not equality of the sexes, but rather the mutual torture that the two sexes inflict on each other if they continue the same old dull round instead of using the imagination to escape or to reimagine existence.

Throughout the poem the narrator's voice is flat, simply narrative, almost deadpan. The only word that the narrator uses to express his own feelings occurs in the third line: "dreadful." Izak Bouwer and Paul McNally interpret this value judgment in a positive way:

> "Men & Women" ... refers to eternal archetypes, and it follows that the poet visited the regions of Man's eternal reality.... The "dreadful" things recounted by the traveler are the events of this land of eternal reality, which are awe-ful, or sublime.[6]

Although this open-minded generosity, in the spirit of Cooke's "fascination" quoted above, may help us to read the poem, such an assertion seems to deny the horror and even the detachment of the narrator. Indeed, most interpreters of the poem see the word "dreadful" as separating the narrator from the events that he describes. However, if, as Gerald Enscoe suggests, the narrator is someone who has already undergone the kinds of experiences that he is about to describe[7], then the narrator might be showing sympathy, not distance. Perhaps, with a vigorous stretching of the imagination, a reader can see the word "dreadful" as all three: positive, pejorative, and sympathetic. The poem thus would set up a complex of attitudes that is not simply either accepting or rejecting, but both.

The "cold Earth wanderers" of line 4 are those of us who fail to view the world with prophetic vision and can not see the horror of what men and women do to each other, and therefore live trapped in these horrors, frozen in our psychological state. "All are bound to the insistent simplicity of a role.... roles exhaust the possibilities of relationship in 'The Mental Traveller.'"[8] The ignorance of which line 4 speaks may not mean that we have not experienced these tortures but that we do not really understand them. Surely every human being who has ever tried to love has lived some of the horrors of this poem and thus "knows" them; just as surely, few of us understand our actions and feelings, and thus we "know" not what we do. By describing the cruel vagaries of love so starkly Blake urges us, as does a satirist, to front the process and thereby find a new way to view the events, perhaps to escape from the horrible cycle by leaving it as does the female babe in the middle of the poem or by transforming the cycle as Blake does in later poems.

Just as the first stanza upsets ordinary categories of men and women, so the second stanza upsets, and even reverses, ordinary categories of sex and birth:

> For there the Babe is born in joy
> That was begotten in dire woe
> Just as we Reap in joy the fruit
> Which we in bitter tears did sow.

Thus, the first incident in the poem is a birth, which in Blake is usually a hopeful sign. According to Martin Nurmi and many other critics, the dreadful cycle that we are about to enter could be broken here, but it is not; the opportunity is missed.[9]

The strange thing about this birth is that it reverses the curse in which birth happens in pain. In the Bible Eve is cursed by God for her transgression of eating the fruit of the forbidden tree:

> I will greatly multiply thy sorrow and thy conception; in sorrow thou shalt bring forth children. (Gen 3:16)

Blake's simple reversal invites us to redefine our unthinking acceptance of the agony of labor, just as the entire poem invites us to reconsider our unthinking acceptance of the battle between the sexes.

In addition to reversing the feeling usually assigned to birth, the poem also reverses the traditional feeling attached to sex, for the babe has been begotten in dire woe. Instead of the expected pleasure in sex and pain in childbirth, the poem gives us pain in sex and pleasure in childbirth. The curses of the Old Testament Urizenic God are thus reversed at the beginning of the poem, but almost every action throughout the poem tries to reinstate them. By reversing traditionally assigned values and feelings, Blake invites us into a potentially transformed world. Each new beginning in the poem promises to reverse an old curse, promises to begin to escape, but then succumbs to the same old dull round.

In this description of the birth of the babe, Blake alludes to a Bible passage in which the Psalmist commemorates the escape from Babylonian captivity:

> When the Lord turned again the captivity of Zion, we were like them that dream. Then was our mouth filled with laughter, and our tongue with singing: then said they among the heathen, the Lord hath done great things for them. The Lord hath done great things for us; whereof we are glad. Turn again our captivity, O Lord, as the streams in the south. They that *sow in tears shall reap in joy*. He that goeth forth ... bearing precious seed, shall doubtless come again with rejoicing, bringing his sheaves with him. (Psalms 126:1–6) [emphasis added]

This allusion implies that a chance for escape from any dreadful condition is possible. Such hints recur throughout the poem, but each such hope gets trampled.

Gerald Enscoe suggests that Blake means that sex has been perverted by Puritanical inhibitions into a dire woe, and that the joy of birth is really the female's relief at ridding herself of her inner burden, which she can now repay for the grief it has caused her.[10] She has been imposed on; now she can impose in turn. Such a reading perfectly catches the cycle of revenge that is set in motion, but it does so at the cost of any genuine joy. Any joy would have to be redefined as sadism. Martin Nurmi, on the other extreme, wants joy to overcome sorrow. He tries to allow both extremes, but can not:

> the emphasis [can be] either on dire woe or on joy. I believe joy to be proper: although the babe is begotten in sorrow, he is born in joy.[11]

I think a combination of these two readings, with neither denying the other, best opens up the Blakean dilemma, unsolved in this poem but solved later in the coincidences of opposites in *Jerusalem*. In "The Mental Traveller," sorrow and joy, stasis and movement, female and male, see-saw back and forth in manic-depressive, sado-masochistic ricochet. Each extreme, by trying to deny the other, by trying to find solidity, as Urizen usually tries to do in *The Book of Urizen* and the *The Four Zoas*, actually forces the cycle to spin faster and more cruelly. The joy and the sorrow are both genuine, but the attempt to destroy either one traps us more inextricably in the dreadful cycles.

As soon as the babe is born, the old woman tries to pin him like a butterfly specimen:

> And if the Babe is born a Boy
> He's given to a Woman Old
> Who nails him down upon a rock
> Catches his shrieks in cups of gold.

When the boy babe is given to the old woman, the first explicit torture of the poem begins. The act can be seen as social; society as a whole allows the torture to happen by giving the child over to the torturer. The passive verb implies a hidden ideology, an action that has no clearly defined actor to be blamed. Haven't we always done it this way? And if we have, how can anyone imagine new possibilities?

The first woman in the poem takes the babe and nails him down upon a rock, thereby trying to force him into rigid, fixed patterns, to solidify him in his state. The imagery recalls Christ, punished by man, and Prometheus, punished by Zeus. This evocation of Prometheus provides interesting echoes into *Prometheus Unbound* and *Frankenstein*, which has as its subtitle "The Modern Prometheus." They both set up a cycle of punishment and revenge which seems inescapable. In *Frankenstein* the cycle is not escaped, but in *Prometheus Unbound,* the power of forgiveness, also the power of liberation in Blake's *Jerusalem,* does break the cycle of revenge.

As in Blake's Orc cycle[12] over and over again, as soon as the spirit of revolution or new life springs up, the forces of repression hasten to pin it down and rigidify it.

> She binds iron thorns around his head
> She pierces both his hands & feet
> She cuts his heart out at his side
> To make it feel both cold & heat.

The iron thorns and the binding of the hands and feet, both reminiscent of the torture of Jesus, represent mental limitations and limitations of physical activity. This kind of imagery, emblematic of the disastrous mutual torture of Ulro, is prevalent in Blake's prophecies. One example, with imagery similar to this image in "The Mental Traveller," emphasizes the feeling of invasion:

> … they cut asunder his inner garments: searching with
> Their cruel fingers for his heart, & there they enter in pomp
> (*Jerusalem* 66:27–28)

Sacrifices abound in Blake's prophecies as examples of the depths of human behavior. Northrop Frye persuasively links sacrifice to the dominance of reason:

> Human sacrifice in all its forms … is the most eloquently symbolic act which the dreaming Selfhood is capable of performing. It illustrates every aspect of the Fall, and parodies every aspect of eternal life.... The motive for human sacrifice is … an effort to express the ascendance of nature and reason in society.[13]

In Frye's explanation Nature and reason band together to torture humanity. Nature, the ideology of keeping what is, provides the excuse for falling into patterns of unimaginative repetition. Reason, the insistence on respectable order, provides the justification for trying to remove recalcitrant or rebellious elements from society. Together Nature and Reason fight to keep out new vision, to restrict the possibilities of life, and to disqualify the coincidence of opposites that is necessary before we can see a way out of the cycle of torture.

As the torture between female and male continues, a cycle of youth and old age sets in. Leaving behind any pretense of realistic travel literature, the poem shows the female moving backward in time:

> Her fingers number every Nerve
> Just as a Miser counts his gold
> She lives upon his shrieks & cries
> And she grows young as he grows old.

In this cannibalistic image, the female grasps, manipulates, and hoards. By living upon the male's shrieks and cries, she implies that she can live only at the expense of his pain. Thus, as in the whole poem, a zero-sum game is played: one contestant can gain only by making the opponent lose. To make herself more and to make the male less, the old woman uses the youthful energy of his protests to nourish herself. In this way new life serves to replenish and nourish old age, but instead of living with the new life in a reciprocal relationship, the old life attempts to control the young life and feed off it vampirishly. Such feeding enables old age to feel that it is not growing older. Old age can not stand to grow older, because it wants to remain in a fixed state in order to fend off death. In a perversion of the Eucharist, the old woman binds the youth and miraculously sucks his blood out. She will not allow him to be an integral being to experience but cuts out his heart and exposes it to try to make him feel the way that she wants him to feel. She forces him to wear his heart on his sleeve, to be sentimental. She wants to destroy him instead of allowing him to grow.

She invades his body even further when she counts every nerve as a miser counts his gold. This image of scientific analysis and hoarding greed reinforces the feeling of control, of misuse. The miser gorges himself on money as the past feeds itself on the present, as the harvester can gorge himself, muttering "I deserve it." This image will crystallize into riches as food later in the poem. The past tries to assure its "futurity" by repressing the present. This perversion can feel certain only if others are tied down and destroyed, or best of all, ingested, assimilated, and negated.

A similar kind of language appears at the beginning of *The Four Zoas* when Tharmas and Enion split apart. Tharmas, who hates and dreads clear articulation, complains:

> Why wilt thou Examine every little fibre of my soul
> Spreading them out before the Sun like Stalks of flax to dry
> The infant joy is beautiful but its anatomy
> Horrible Ghast & Deadly nought shalt thou find in it
> But Death Despair & Everlasting brooding Melancholy.
> (*Four Zoas* 4:28–33, E298)

The analysis that Enion and the old/young woman perform on the males takes vivid physical form, but it arises from rationalistic analysis, the kind that wants to take everything apart and see what makes it tick. "Our meddling intellect" does "murder to dissect."

When Urizen abjures his error near the end of *The Four Zoas*, he relinquishes his desire for a fixed futurity:

> O that I had never drank the wine nor eat the bread
> Of dark mortality nor cast my view into futurity nor turnd
> My back darkning the present clouding with a cloud
> . . .
> Seeking the Eternal which is always present to the wise
> Then Go O dark futurity I will cast thee forth from these
> Heavens of my brain nor will I look upon futurity more
> I cast futurity away & turn my back upon that void
> Which I have made for lo futurity is in this moment
> (E390)

Urizen's repentence is one of the solutions to the problem set up in "The Mental Traveller." That solution never is stated in this poem, but it is always potential.

As the cycle goes on, the female grows younger and the male grows older until they reach the same age:

> Till he becomes a bleeding youth
> And she becomes a Virgin bright
> Then he rends up his Manacles
> And binds her down for his delight

The opportunity for mutuality appears as they reach same age, but the moment of possible renewal passes as fast as it arrives, and the male reciprocates the tortures. As with the births of the babes at the start and end of the poem, the moment of potential freedom becomes instead an opportunity for further exploitation. Like the cycle of revenge that ruins the newness of birth, the perverted reciprocity of revenge holds on to the previous condition and allows it to infect the present state. At the birth of the babe, the female ruined the potential for escape by binding him; now at the conjunction of ages, the male ruins the potential for escape by binding her. Any attempt to hold on to the past is an attempt to freeze time, to allow no further dissolution of an egotistical self that hates change. And holding on to vengeance from the past turns the screw even further.

In Enscoe's Freudian reading of the male and female as son and mother,[14] the mother seems to grow younger as the boy grows older and she seems to become just another woman. In Enscoe's reading, just as the woman finds joy in punishing that which gave her sorrow, the male breaks his chains and ties her down for his delight. Each revenges previous pain. The cycle of revenge goes on. The potential for breaking out of the cycle is destroyed by the binding which gives the male only a kind of perverted pleasure, based, in Frye's terms, on jealousy:

> The abstract reasoner cannot see a tree without dragging its shadow off to the cave of his own mind.... The Selfhood cannot love in the sense of establishing a kinship with the beloved: it can regard the latter only as a possession, something to contemplate in solitude.[15]

When the male invades the female's nerves as she had invaded his, the torture becomes even worse. The cycle does not just repeat itself; it spirals into more intense torture. Whereas she counted and anatomized, he goes further by planting himself, becoming a part of her, invading her more deeply than she did him:

> He plants himself in all her Nerves
> Just as a Husbandman his mould
> And she becomes his dwelling place
> And Garden fruitful seventy fold

Now instead of metaphors of divine and human sacrifice, the metaphor becomes one of gardening. The female becomes mold or fertile earth, to be planted in by the seed of the male. Although plant and garden imagery is often pejorative in Blake, in some cases it does herald a possibility of escape from disaster:

> in Beulah the Feminine
> Emanations Create Space. the Masculine Create Time, & plant
> The Seeds of beauty in the Space
> (*Jerusalem* 85:7–9)

Near the end of *Jerusalem*, in the most important Blakean example of the redemptive value of gardening, Los realizes that he must act not only as a hammering, active blacksmith, but also as a waiting, passive farmer:

> The land is markd for desolation & unless we plant
> The seeds of Cities & of Villages in the Human bosom
> Albion must be a rock of blood. (83:54–56)

Again, however, the cycles of "The Mental Traveller" blast the potential for fulfillment. Although the garden image contains hope, the male character instead exploits its vengeful portion: by domesticating the female, he makes her into the ultimate housewife—a wife who is also a house: "And she becomes his dwelling place." Some interpreters, such as Nurmi and Bouwer, see this stanza as positive:

> The Spiritual principle [male], now dominant, is able to control the natural world [female] with increasing ease and joy, where before ... it could express itself in nature only through suffering.[16]

Yet Enscoe pinpoints both the extremity of the male's invasion and its potential for reversal. He calls the action,

> procreation by domination. It is 'himself' he is planting, and ironically the self he plants will become the female Babe who will drive him away into the desert later in the poem.[17]

Thus Enscoe's reading incorporates the favorable interpretation by Bouwer, but judges the male's planting as cruel, even while pointing out its unexpected, freeing result. At this point, a reader sees the diminishment of the female, but can hardly imagine her imminent liberation.

Soon the female, who has been so degraded, seems to disappear from the poem as the old man wanders around the house all alone:

An aged Shadow soon he fades
Wandring round an Earthly Cot
Full filled all with gems & gold
Which he by industry had got

In their insistence on maintaining a positive view of the male, who in their reading consistently represents the human spirit as against the female world of nature, Bouwer and McNally interpret this stanza as a positive view of the male. Thus, by trying to maintain a consistent, fixed view, they fall into laborious distortions:

> We suggest that "aged" merely indicates that the Spritual principle is nearing full manifestation and greatest potency.[18]

The male has achieved a very perverse potency indeed: not only has the woman been reduced to garden and house, but she has been reduced to smaller, precious commodities—jewels. When a woman is referred to as a jewel, she is being equated with the object of greed, miserliness, and possession. She is small, beautiful, helpless, in fact almost non-existent, except as an embodiment of the ultimate object of man's desire.

The gems become emblems simultaneously of courtly love and religious suffering:

And these are the gems of the Human Soul
The rubies & pearls of a lovesick eye
The countless gold of the akeing heart
The martyrs groan & the lovers sigh.

In this stanza Blake rings changes on his own short poem, "Riches":

The countless gold of a merry heart
The rubies & pearls of a loving eye
The indolent never can bring to the mart
Nor the secret hoard up in his treasury.
(E461)

These gems of love and joy in this other poem can not be trapped and hoarded. But the twists of pain and jealousy can become the miser's object. Blake delineates the psychology of this dynamic in one of the most bitter passages in *Jerusalem*:

> All Quarrels arise from Reasoning. the secret Murder, and
> The violent Man-slaughter. these are the Spectres double Cave
> The Sexual Death living on accusation of Sin & Judgment
> To freeze Love & Innocence into the gold & silver of the
> Merchant
> Without Forgiveness of Sin Love is Itself Eternal Death.
> (64: 20–24)

The gems thus, like most of the images in the poem, contain a dual potential: they can be pain or they can be joy. If joy, they are free and creative. If pain, they are trapped and unimaginative, but still bursting with potential. Like every other potential in the poem, though, they are twisted into exploitation. As the ultimate exploited object, they become food:

> They are his meat they are his drink
> He feeds the Beggar & the Poor
> And the wayfaring Traveller
> For ever open is his door.

Consistently, Bouwer and McNally interpret this image as favorable:

> Stanzas 10 and 11 ... describe the state of Eden.... the eternal forms appear fully assembled as the 'Family Divine.' This family is the 'Council of God.'[19]

Although exaggerated, this interpretation does underline the potential of regeneration. Greediness has become charity, at least, although Blake's comments on charity in other contexts make the value ambiguous at best. When the woman was old and the man was young, she fed off him. Now the positions are reversed as the cycle has turned almost 180 degrees. Just as they torture and submit to each other in perversions of genuine love, so the aged man takes the profits that he has made by exploiting the woman, does not even leave them in their organic form, but hardens them, makes them metallic, and uses them to exploit others in the name of charity. "Pity would be no more if we did not make somebody poor." And indeed the old man has already done his part to make another poor by exploiting the woman earlier. His grief, like the devouring joy in stanza 1, makes others happy:

> His grief is their eternal joy
> They make the roofs & walls to ring
> Till from the fire on the hearth
> A little Female Babe does spring
>
> And she is all of solid fire
> And gems & gold that none his hand
> Dares stretch to touch her Baby form
> Or wrap her in his swaddling-band.

Now the female reappears. Having been reduced into complete domesticity, into becoming a very house, she now is reborn from the hearth, the ultimate epitome of the household. The female, who had seemingly been eliminated, thus reveals herself as a principle that can not be destroyed. She cannot be repressed. She must return. The apparent reduction, even destruction of the female has not taken place at all. She has merely changed her conditions, her states, and now comes back stronger than ever.

According to Morton Paley, the female babe is the archetypal evil female, who

> unites numerous evil females of Blake's pantheon.... She is the sum of what the male has created up to this point and so represents, not an imaginative achievement at all, as some have suggested, but the entirely materialist values of the urizenic Host.[20]

According to Adams, at the other extreme, she is a force for creativity. He even goes so far as to assert that the man she loves is the narrator with whom she leaves the poem.[21]

Once again, a reader who can see both these extremes at the same time comes closer than the reader who can see only one extreme. Of course Paley is correct that the female babe is a result of the materialistic values of the host; after all, she has been beaten down and domesticated by him. On the other hand, Adams is correct that she is a counter force to materialism, that she is a creative force. For Paley assumes that no creativity can arise from its diametrical opposite, materialism. But the point of Blake's use of the patterns of *coincidentia oppositorum* is that creativity can and does arise from materialism. Paley's vision would miss that miracle, and would probably not understand the potential for renewal at this mid-way point in the poem. Such a miss would allow the horrible cycle to continue while the female babe is about to escape from tyranny into freedom.

Enscoe's interpretation accommodates the doubleness, but goes to the extreme of missing the negative aspects of exploitation:

> This grief ["His grief is their eternal joy"] is positive. If the Host felt no grief, if he were happy in his position as domineering tyrant, no hope of change would exist, no movement would take place from one 'state' to another, no "wanderers" would ever become "travellers."[22]

While Enscoe is correct in pointing out that the grief produces changes in state that may lead to a more comprehensive vision, and while his insight that some of the "cold Earth wanderers" of line 4 may become mental travellers with the broad vision of the narrator gives a valid hope, he neglects to acknowledge the genuine pain of materialism and exploitation. The gems to which the female had been reduced now reappear as the solid gems of the female babe, active instead of passive, strong instead of weak, free instead of imprisoned. Thus the limit of reduction turns back against its oppressor, but by refusing to repress him in turn, escapes from the trap, in a way similar to the escape of Prometheus at the start of Shelley's *Prometheus Unbound*.

We are not told why she can not be touched but the strong implication is that she is simply too fierce. She resists the ideology of naturalness that allows the boy babe in stanza 2 easily to be handed over to the old torturing woman. I agree with Enscoe's reading that the female babe is a hopeful sign,[23] the most hopeful among many such moments in the poem, similar to but even stronger than the birth of the young male at the beginning of the poem, and his reappearance at the end of the poem. The young female is a sign that there is hope to break from the cycle. She is the only one who does not get caught in the cycle. And indeed she will be the only character who does escape from the cycle to find a new life. Every new birth gives such a hope, whereas the mere cycle of young-old-young-old-young promises nothing but the same old mistakes over and over again. Not only can no one touch her, but no one can wrap her in a swaddling band. This image not only reminds us that she cannot be restricted, but also reminds us of the birth of Christ, the sacrifice par excellence, who was invoked in stanzas 3 and 4. She is neither a passive sexual victim nor a sadistic sexual aggressor, as is every other character in the poem.

> But She comes to the Man she loves
> If young or old or rich or poor
> They soon drive out the aged Host
> A Beggar at anothers door.

She uses the power of her will to make a conscious choice instead of imposing herself or being imposed upon, as do the other characters. "The female babe has choices beyond the perceptual scope of the poem."[24] And she does not use the state of the man, "young or old or rich or poor," to determine her choice. Instead of acting out of a reflex of revenge, she, as an individual, chooses another individual rather than simply reacting to him because of his state. She freely comes to the man she loves without regard to his condition because she will not exploit him, and she will not be exploited by him. Bloom reads her choice of a lover as a rejection of age in favor of youth.[25] But such an interpretation flies in the face of the explict disclaimer, "if young or old or rich or poor." The categories that determine the other actions in the poem do not determine hers, just as the ideologies of "nature" and reason do not determine the decisions of Blake.

She is not choosing youth, but a different psychological state, just as her original condition after birth was a state not amenable to the tortures of the cycle. What she chooses is so different from the values of the rest of the poem that she must disappear from the poem entirely. What is important is that she freely chooses her lover and makes a new life for herself outside the cycle. Many critics, including Bloom, see this as nature turning against man, but in fact the man carries on the cycle without her because she has left the cycle. She did not turn against him. She refused the cycle of mutual torture. We see her no more because she now lives beyond this poem, in a different state, a different dimension, which we will not see directly until the conclusions of Blake's epics.

This frozen man's own frozen state drives him out. He is driven out because he deserves to be. Since above I have used one of Blake's definitions of the prophetic character as a general guide to my interpretation of "The Mental Traveller," one of his other definitions of the prophet will be helpful at this point. The prophet, according to Blake, is not one who can simply predict the future. The prophet, speaking his opinion frankly, is one who sees deeply into the meanings and results of people's actions:

> Every honest man is a Prophet he utters his opinion both of private & public matters/Thus/If you go on So/the result is So/He never says such a thing shall happen let you do what you will.[26]

The entire cycle of "The Mental Traveller" is caused by the ways in which the male and female go on. The old man has not been harmed by the female babe. Quite the contrary, he has created the very situation in which he must be driven out into the freezing cold. Clinging, reducing, trying greedily to grasp his present state, he has created his own metaphysics, as does the angel in *The Marriage of Heaven and Hell,* who sees the horrors of spiders, Leviathan, and tygers, and is told, "All that we saw was owing to your metaphysics" (E42).

Having exploited the woman-garden, hoarded up his jewels like a miser, and then pityingly fed off them and fed them to others, the old man loses the love of the female babe, who will not be bound. He has reached the limit of contraction; he has pushed so far that he gets pushed back. The old man who pitied and fed the beggars now becomes himself a beggar; he becomes what he beheld. Whereas earlier he wandered inside his own cottage, this time his wandering takes him far away. At least in his cottage surrounded by his jewels, he could feel some material security. Now completely lost, his security fades even further away. Miserliness and jealous holding do not guarantee that he can keep his possessions; despite (or perhaps because of) his greed, his possessions flee his grasp. He is now pitiful himself and dependent on the mercy and charity of others:

> He wanders weeping far away
> Untill some other take him in
> Oft blind & age-bent sore distrest
> Untill he can a Maiden win

Wandering and lost, he thinks his distress can be alleviated only by winning, conquering a maiden. He wants her as an object because he can not live with an independent female. Although the woman he finds is not a continuation of the female babe, who has chosen a path out of the cycle, she does seem to be a continuation of the earlier females who were bound down. Although he is wandering in time and space, he remains in the same psychological state of egotistical grasping. When the female babe breaks out of the cycle, the male keeps the cycle going by finding a substitute. Refusing the potential for escape that he has just been shown, he tries to recapture the domestic woman of line 27. He does not come to her or allow her to come to him, but instead insists on winning her as a prize, thus perpetuating the cycle of exploitation. Instead of two mutually agreeing to come together, one dominates the other, making the other into an object of torture and perverse pleasure:

And to allay his freezing Age
The Poor Man takes her in his arms
The Cottage fades before his sight
The Garden and its lovely Charms.

Once he grasps the maiden, even the little vision that remained fades away. He has reached his limit of age, his limit of contraction, and now even his false visions disappear. He has already been forced (because he was first an exploiter and then a pitying philanthropist) to leave the cottage, garden, and inn and wander far away from them. When he tries to regain what he has lost, even the re-won cottage and garden fade from his sight. Even the possibilities of the kind of perverted love that he possessed before have disappeared. His grasping exploitation destroys life, and so now the whole universe becomes a desert which his perversions have created for him. The cycle worsens: instead of a garden and jewels on which to feed, he has no food at all. The guests who had been able to find some kind of shelter and food are now scattered, just as the old man himself had previously been driven away:

The Guests are scatterd thro' the land
For the Eye altering alters all
The Senses roll themselves in fear
And the flat Earth becomes a Ball.

The last two lines of this stanza, although not simple, are fairly easily understood: Blake consistently prefers a vision of a flat Earth to that of a round one. The round Earth is a result of the roundness of the eye, which is a limitation of the potential for infinite vision. The second line of the stanza, however, suggests Blake's peculiar delineations of perspectives of vision. Those perspectives range from a rather simple concept—our perceptions are limited to our immediate surroundings—to a complex conceit of vortices that strains even the best interpretive talents of Northrop Frye and Harold Bloom.

In a strange (but strangely commonsensical) passage in *Milton*, Blake redefines a person's universe as the space in which he lives:

> The Sky is an immortal Tent built by the Sons of Los
> And every Space that a Man views around his dwelling-place:
> Standing on his own roof, or in his garden on a mount
> Of twenty-five cubits in height, such space is his Universe;
> And on its verge the Sun rises & sets. the Clouds bow
> To meet the flat Earth & the Sea in such an orderd Space:
> The Starry heavens reach no further but here bend and set
> On all sides & the two Poles turn on their valves of gold:
> And if he move his dwelling-place, his heavens also move.
> (*Milton* 29:4–12; E126)

A reader might think that Blake is mocking this limited kind of existence, in which the individual perceives only his own tiny world. But as the passage goes on, Blake attacks abstract cosmography:

> Where'er he goes & all his neighbourhood bewail his loss:
> Such are the Spaces called Earth & such its dimension:
> As to that false appearance which appears to the reasoner,
> As of a Globe rolling thro Voidness, it is a delusion of Ulro.
> (*Milton* 29:13–16; E126)

The reader of *Milton* is given much help in that poem to grasp this concept of space: instead of an abstract round globe that can be conceived only by reason, a globe that no one can perceive, Blake gives us a homey neighborhood of a universe, individualized, mobile, and alive. We do not receive much help with this concept in "The Mental Traveller," and yet it seems essentially the same: each individual carries a universe, which can be changed if the individual's perceptions change, for better or for worse.

So far, so good. One of Blake's most notorious passages, however, ratchets this issue of perspectives up a few notches:

> The nature of infinity is this: That every thing has its
> Own Vortex; and when once a traveller thro' Eternity
> Has passd that Vortex, he percieves it roll backward behind
> His path, into a globe itself infolding; like a sun:
> Or like a moon, or like a universe of starry majesty,
> While he keeps onwards in his wondrous journey on the earth
> Or like a human form, a friend with whom he livd benevolent.
> As the eye of man views both the east & west encompassing
> Its vortex; and the north & south, with all their starry host;
> Also the rising sun & setting moon he views surrounding
> His corn-fields and the valleys of five hundred acres square.
> Thus is the earth one infinite plane, and not as apparent
> To the weak traveller confin'd beneath the moony shade.
> Thus is the heaven a vortex passd already, and the earth
> A vortex not yet pass'd by the traveller thro' Eternity.
> (*Milton* 15:21–35)

Nurmi uses this passage as a way of explaining what happens in "The Mental Traveller":

> The aged outcast passes through a 'vortex'—that curious optical-symbolic phenomenon of Blake's in which a perceiver goes as far with one kind of perception as he can and passes through the object, as it were, to the other side to a different way of looking at it.[27]

This solid explanation clearly implies the reversal in vision that happens in the poem: flat planes become spheres as the perceiver seems to be seeing through the wrong end of a weird telescope.

Harold Bloom's explanation captures the doubleness of eternal vision as he puts center and circumference together:

> The vortex is the eddy or whirlpool of eternal consciousness, whose center is the object eternal consciousness intends. Since center and circumference are not separate in eternal vision, the perceiver is at once the apex of his vision, and yet able to regard it from a distance. But when Milton

> passes into Beulah, he leaves eternity for time, and moves to the apex of his own vision. He is thus objectified, and the eternal circumference of his vision rolls up behind him. The eddy of perception is solidified into the globed universe of Newtonian observation. What survives of eternal vision depends upon the temporal perceiver's imagination, for he can still encompass his vortex and see the object world in its human dimension (line 27), as "one infinite plane" (32).[28]

Bloom's explanation may be more obscure than Blake's passage. As so often, the best explanation comes from Northrop Frye:

> ... when we see ourselves as imprisoned in a huge concave vault of sky we are seeing from the point of view of a head that is imprisoned in a concave vault of bone.... Blake says that everything in eternity has what he calls a 'vortex' (perhaps rather a vortex-ring), a spiral or cone of existence. When we focus both eyes on one object, say a book, we create an angle of vision opening into our minds with the apex pointing away from us. The book therefore has a vortex of existence opening into its mental reality within our minds. When Milton descends from eternity to time, he finds that he has to pass through the apex of his cone of eternal vision, which is like trying to see a book from the book's point of view; the Lockian conception of the real book as outside the mind on which the vision of the fallen world is based. This turns him inside out, and from his new perspective the cone rolls back and away from him in the form of a globe. That is why we are surrounded with a universe of remote globes, and are unable to see that the earth is 'one infinite plane.' But in eternity the perceiving mind or body is omnipresent, and hence these globes in eternity are inside that body.
>
> Before the Fall, Man was abolute wisdom, and was the circumference of everything. Nothing then existed outside Albion: sun, moon, stars, the center of the earth and the depth of the sea, were all within his mind and body, a body fully conscious of being alive, not only in its brain, but in all parts of itself down to the feet. Hence 'opening a centre,' as described above, is the imagination's way of revers-

> ing the fallen perspective of the world, and uniting an individual imagination with the universal one.[29]

Of course the process in "The Mental Traveller" is the reverse of the kind of opening that Frye describes. Instead of achieving a visionary perspective, the male in the poem is making sure that he deepens his fallen perspective.

The reason that the narrator can see what cold earth wanderers can not is that he has adjusted his vision so that he sees the opposites of center and circumference simultaneously. He has passed through the states of pain and jealousy that pervade that poem. But the characters within the poem, with the possible exception of the female babe and her lover, do not see clearly and in fact do not take the opportunities offered them to see clearly. Instead they perpetuate the cycle of repression, torment, and submission.

The loss of vision and sustenance continues as vision continues to shrink:

> The Stars Sun Moon all shrink away
> A desart vast without a bound
> And nothing left to eat or drink
> And a dark desart all around.

The Ptolemaic universe, become the Copernican, is now the Newtonian, where all is mathematical and finite. Instead of an infinite flat earth close to the heavenly bodies, the altered vision of the aged man creates a sterile, mathematical earth, infertile with nothing but desert. At least earlier he had been able to plant a garden, even if it was built on the pain of another. Instead of a vision that includes the coincidence of opposites, this vision can see only separation.

Again, it is important to emphasize that the poem itself does not give us direct reason for hopes of fuller vision, but in juxtaposition to those poems that do, we can obtain a fuller understanding of the limits in "The Mental Traveller."

Now at the limit of sterility, the male figure begins to return to his childhood.

> The honey of her Infant lips
> The bread & wine of her sweet smile
> The wild game of her roving Eye
> Does him to Infancy beguile.

The honey seems like the manna in the desert for the children of Israel in their wanderings. The bread and wine seem like the Eucharist. However, the imagery turns sour as the stanza proceeds. The honey and bread and wine become a wild game from a roving eye. Flirtation keeps the cycle going. As a corollary to the old man who could imagine only a maiden whom he could win and clutch, now the female can imagine only a man whom she can lead on, instead of meeting directly. She does not simply lead or inspire him but beguiles him, with more than a hint of guile.

For as he eats & drinks he grows
Younger & younger every day
And on the desart wild they both
Wander in terror & dismay.

Again a moment of escape seems to flit past as the male grows younger to reach the age of the female. In stanza 6 the terms "bleeding youth" and "virgin bright" clearly implied that they were about the same age. Here in stanza 19 the implication is more subtle, but as the strange ascent and descent in age continues, they must once again be at the same age. Missing their opportunity, they struggle on, lost and separate.

The splitting of male and female is predominant in Blake's myth, usually based on the Bible creation story, and usually personified in Los and Enitharmon:

She separated stood before him a lovely Female weeping
Even Enitharmon separated outside, & his Loins closed
And heal'd after the separation: his pains he soon forgot:
Lured by her beauty outside of himself in shadowy grief.
Two Wills they had; Two Intellects; & not as in times of old.

Silent they wanderd hand in hand like two Infants wandring
From Enion in the desarts, terrified at each others beauty
Envying each other yet desiring, in all devouring Love
(*Jerusalem* 86:57–64)

He feeds off her, just as earlier the man had made the woman into his garden. But this time the female is the active one, he the willing victim. In perversions of potential sexual relationships the female feeds off the pain of the male, and the male feeds off the teasing of the female. The intensification of the horrors of the cycle leaves the two main characters now wandering together and yet apart in the desert. No longer do we have a rock with nails, nor do we have a cottage or a garden, all with some kind of permanence and security. For security can not be forced, and the more one tries to force it, the less one has it.

> Like the wild Stag she flees away
> Her fear plants many a thicket wild
> While he pursues her night & day
> By various arts of Love beguild.

Now instead of her active teasing, her passive fear is emphasized. The result of that fear is a planting of thickets, obstacles to the male's pursuit. And yet soon those thickets change their connotation from obstruction to fertility in the desert:

> By various arts of Love & Hate
> Till the wide desart planted oer
> With Labyrinths of wayward Love
> Where roams the Lion Wolf & Boar.

The second line of that stanza again gives us a glimpse of a possibility of fertility, but soon that possibility is perverted into messy "Labyrinths." In a technique similar to that in the opening two lines of the poem—"I travelled through a land of Men / A Land of Men & Women too"—Blake repeats the last line of stanza 20 in the first of 21: "By various arts of Love beguild / By various arts of Love and Hate." The repetition with addition reveals a split concealed by the traditional term. As generic man includes hidden woman in the first stanza, so generic love includes hidden hate. The most basic opposites in human existence, as many clichés realize, live inextricably together.

When the male reaches his renewed youth and the woman her renewed old age, we feel that we are very close to where we started. In fact we are almost there, but before we get to that terrible promise of repetition of the cycle, we see one more glimmer of hope, the most beautiful and extended image in the poem:

> Till he becomes a wayward Babe
> And she a weeping Woman Old
> Then many a Lover wanders here
> The Sun & Stars are nearer rolld
>
> The trees bring forth sweet Extacy
> To all who in the desart roam

Till many a City there is Built
And many a pleasant shepherds home.

Many interpreters see this passage as positive, as another chance for renewal:

> The moment when man may return to Eden is at hand, and everything is fruitful and productive. A Second Coming has arrived, if only mankind knew it.[30]

> While the male is still a Babe, and the woman is merely weeping over the past but is not actively continuing the cycle, a brief glimpse of another possibility is presented.[31]

The world that had shrunk and fallen away from the stars, the world that had become a desert without civilization or shepherds, has now become a city and a pasture once again. The labyrinths of thickets caused by the torments of love and jealousy in stanzas 20–21 have been transformed into sweet trees. The beauty of this scene offers itself for man and woman if they will only accept it. The two people are not happy, to be sure, for he is wayward and she is weeping, but there is hope of possible repentance here. After all, hasn't the altering eye provided its own best environment? Many lovers wander here how; even if lost, they may hope that the guests will return, that the mistakes of the past can be made up for.

Suddenly, after this respite, the horrors of the cycle reassert themselves. There is no joy in the discovery of the babe this time, as there was in stanza 2. In fact the babe is not "born" as he was in the beginning of the poem, marking a new beginning. Instead he has reached his present condition by regressing from adulthood.

But when they find the frowning Babe
Terror strikes thro the region wide
They cry the Babe the Babe is Born
And flee away on Every side

For who dare touch the frowning form
His arm is witherd to its root
Lions Boars Wolves all howling flee
And every Tree does shed its fruit

And none can touch that frowning form
Except it be a Woman Old

She nails him down upon the Rock
And all is done as I have told.

In the beginning the babe was easily given to the old woman, but here he can be touched by none but an old woman. Anyone else who tries to touch (in order to bind) destroys himself. Even the wild beasts that helped make the labyrinths of wayward love more terrifying are terrified by this new babe. The trees that had brought forth ecstasy give up their fruit.

The last line of the poem—"And all is done as I have told"—seems to indicate that the cycle will be repeated endlessly. In fact, since the babe seems even more terrifying than he did in stanza 2, the ending implies that the cycle will progressively worsen. And yet, despite its apparent bleakness, the poem works to encourage in us the hope that the horrible cycle can be broken. In fact, if the babe can not be so easily rounded up, perhaps he, like the female babe in the middle of the poem, can find a new consciousness outside this closed universe.

Two critics in particular enunciate a hopeful interpretation of the poem. Enscoe and Adams both place their faith in the voice of the narrator. Enscoe claims—and his claim I believe can be allowed only on the strength of the first stanza—that the narrator clearly speaks from a larger perspective:

> Blake has presented an alternative to this world of the male-female struggle for domination.... a voice speaking from a state that allows men to see beyond present reality.[32]

Adams goes even further to link the narrator with Blake's creative hero in the later poems and indeed to implicate Los in the poem itself: "The speaker of the poem is Los and he has been loved by the female babe."[33] Thus the narrator has actively participated in breaking the cycle by escaping with the female babe.

I think that both Enscoe and Adams read too much into the poem, but I do agree with the spirit of their interpretations. Although the poem itself does not provide us with such a robust hope, we can legimately read it in the context of Blake's later works, where such robust hope unmistakably appears. "The Mental Traveller," then, is the scorching, despairing portrait of human existence, epitomized in male-female struggles, when caught in the ratio of limited vision. It is all the more painful in that the characters constantly miss the glimpses of a larger, more poetic vision that would allow them to escape the cycle.

As the cycle spins, centrifugal force drives the characters out. Only the female babe takes the leap provided by such propulsion. All the other characters impose the centripetal force of clinging egotism, trying to hold the sacred, secret center. Instead of creating a dynamic interpenetration of opposites, they create a deadly balance.

Notes

1. *Acts of Inclusion* (New Haven: Yale UP, 1979), p. 150.

2. *Wheels of Eternity* (NY: St. Martin's, 1990), pp. 3, 101.

3. *The New Apocalypse*, (Lansing: Michigan UP, 1967), pp. 215–16.

4. Northrop Frye, *Fearful Symmetry* (Princeton UP, 1969), p. 221.

5. *Hamlet* II.ii.295.

6. Izak Bouwer and Paul McNally, "'The Mental Traveller': Man's Eternal Journey," *Blake: An Illustrated Newsletter*, Winter 1978–79, p. 186.

7. Gerald Enscoe, "The Content of Vision," *Papers on Language and Literature*, 4 (1968), p. 403.

8. Cooke, *Acts*, p. 153.

9. Martin Nurmi, "Joy, Love, and Innocence in Blake's 'The Mental Traveller,'" *Studies in Romanticism*, 3 (1964), p. 112.

10. Enscoe, p. 405.

11. Nurmi, p. 110.

12. Frye, *Fearful*, pp. 207–35; Nurmi, p. 109.

13. Frye, *Fearful*, pp. 397, 399.

14. Enscoe, p. 405.

15. Frye, *Fearful*, p. 72.

16. Bouwer, p. 187.

17. Enscoe, p. 406.

18. Bouwer, p. 187.

19. Bouwer, p. 187.

20. Morton Paley, "The Female Babe and 'The Mental Traveller,'" *Studies in Romanticism*, 1 (1962), p. 97.

21. Hazard Adams, "The Mental Traveller," in Adams, *William Blake: A Reading of the Shorter Poems* (Seattle: Univ. of Washington, 1963), p. 100.

22. Enscoe, p. 408.

23. Enscoe, p. 408.

24. Donald Ault, *Visionary Physics* (Univ. of Chicago Press, 1974), p. 189.

25. *Blake's Apocalypse* (Ithaca: Cornell UP, 1963), p. 294.

26. "Annotations to Watson," E617.

27. Nurmi, p. 114.

28. Bloom's commentary, E829.

29. Frye, *Fearful*, p. 350.

30. Nurmi, p. 115.

31. Enscoe, p. 412.

32. Enscoe, p. 413.

33. Adams, "Mental," p. 100.

Chapter 7
Blake's Internal Eternity: Self Becomes Other

Like Coleridge, Blake will settle for nothing less than all. David Erdman reports in the preface to his Concordance to Blake's writings that "all" is the most frequent word in Blake's writings, more than twice as frequent as the next word (except for the commonest English words such as "and," "the," "he," etc.).[1] The similarity between Blake and Coleridge was so striking to some observers that the *London University Magazine* reported:

> A witness to a meeting of the two reported that 'Blake and Coleridge, when in company, seemed like congenial beings of another sphere, breathing for a while on our earth: which may be perceived from the similarity of thought pervading their works.'[2]

Even Coleridge recognized the similarity, but gave pride of place to Blake in the realm of the other world:

> You perhaps smile at [my] calling another poet a [Mystic]; but verily I am in the very mire of common-place common-sense compared with Mr. Blake, apo- or rather—anacalyptic Poet, and Painter![3]

Without mentioning Blake, Richard Holmes describes Coleridge in very Blakean terms:

> Coleridge's own imagination belongs to a distinct literary tradition: it is deeply English, rural, and with a strong idealising or neo-Platonic strain.... Everywhere it seeks the 'radiance' of the eternal in the particular.[4]

Yet, while Blake and Coleridge may work from very much the same presumptions and predilections, predilections which help them lean toward the coincidence of opposites, Coleridge eventually becomes a conservative, using the coincidence of opposites to support the status quo. Blake, however, uses the coincidence of opposites to burst bounds, to redefine all of existence. After the disastrous, inescapable cycles of "The Mental Traveller," his epics, especially *Jerusalem*, reveal a powerful use of the coincidence of opposites to achieve a more imaginative plane of existence:

> For Blake ... Jesus the Imagination, rather than taking part in a Coleridgean unification and idealization, is an iconoclast....[5]

That iconoclasm exhibits itself early in *Jerusalem* when Blake states the purpose of his work:

> To open the Eternal Worlds, to open the immortal Eyes
> Of Man inwards into the Worlds of Thought: into Eternity
> Ever expanding in the Bosom of God, the Human Imagination.
> (5:18–20; E147)

The movement into Eternity is both interior and exterior, penetrating into depths and expanding into heights. Like the writers already discussed, Blake sets up circles of expansion and contraction, but instead of presenting them as contrary forces in balance or reconciliation or oscillation, he boldly assumes that they are identical. The centripetal, inward force that penetrates into "the Worlds of Thought" enables the poet to "open the Eternal Worlds," juxtaposed in Blake's syntax to an identical, centrifugal force, which is "expanding in the Bosom of God." Instead of the philosophical arguments of a Cusanus or a Pope or a Coleridge, instead of the agnonized tensions of a Mary Shelley or a Percy Shelley or the Blake of "The Mental Traveller," the Blake of *Jerusalem* simply states his outrageous coincidence of opposites in the most declarative terms. Furthermore, he conflates God and Man without any of the geometrical inventiveness of Cusanus or any of the soul searching of Coleridge. In simple geometrical terms, the expanding circumference of God equals the focussing center of humanity.

Several critics imply, in various ways, that Blake's inward movement emphasizes the inward at the expense of the outward. Otto indicts most of us:

> Blake criticism, particularly since the work of Northrop Frye, has worked within a discourse which tends to erase the very distinction between self and other, and time and Eternity. As a result the question of how our worlds are to be opened, and how we can perceive what is other, does not appear in its full force.[6]

Attention to the concept of the coincidence of opposites can help to rectify that distortion. Whereas Cusanus, Pope, Coleridge, Mary Shelley, and Percy Shelley find balance or unbearable tension or unsolved mystery in the oppositions of inward and outward forces, Blake plays in both forces to the full extent of their power. His "The Mental Traveller," explored in the previous chapter, shows what can happen when the opposing forces merely try to counter and even dominate each other: they create circles of torture. His later works, especially *Jerusalem*, show how one can escape from the horrible circles, essentially by plunging more deeply into the opposing forces and allowing them to interpenetrate.

Stephen Behrendt, like Otto, emphasizes the outward rather than the inward:

> Blake's own art is explosive rather than implosive in its intellectual and aesthetic signals, directing its audience outward even when it appears to be doing just the reverse: the objective is to see not so much the grain of sand as the World it contains, to 'Hold Infinity in the palm of your hand / And Eternity in an hour'[7]

Behrendt's explanation is useful as a corrective to too much emphasis on the inward, but it should not be read as an absolute in which explosion replaces implosion; both are necessary.

I want to concentrate my discussion of Blake's coincidences of opposites in *Jerusalem* especially on his interpretation of the Bible, which, according to Northrop Frye, is one of the only two approaches to the poem:

> In reading *Jerusalem* there are only two questions to consider: how Blake interpreted the Bible, and how he placed that interpretation in an English context.[8]

I shall neglect the latter admonition to pursue the former. In *Jerusalem* Blake uses the Bible to teach his readers to look inward and expand outward, a simultaneous double movement that reveals Eternity. On plate 16 the Bible is equated with the sculptures in Los's Halls where every possible story is told. From that point of view the Bible is a revelation, hammered out in detail by the artist. On plate 48 the Bible makes up the pillars of the couch where Albion awaits the resurrection. From that point of view, it is a creation of mercy, given to man to keep him safe. But in this latter scene, the couch which is identical with the Bible is brought to man by his enemies, as well as by Jesus, and thus appears also as a tomb. Therefore the Bible is both evidence of our fallen condition and at the same time a solace and a means of escape from the fallen condition.[9]

Blake insists that human perceptions and actions must be raised from the fallen, temporal world which seems all too obvious to men's eyes into the eternal world of Vision which he is trying to open to them, and open them to. Paradoxically the only way to reject that fallen world is to embrace it; the eternal and the temporal are inseparable, even though they are exact opposites. Fallen vision refuses to see this fundamental identity and tries to separate the two worlds, either mystifying or ignoring the Visionary world. Like the opposing forces in "The Mental Traveller" fallen vision tries to divide the contraries (in that poem represented by male and female) and pit them against each other because they believe that they must destroy or dominate each other instead of seeing the other as identical to the self even while other than the self. Eternal Vision reverses and then includes fallen vision by forgiving and transfiguring it. For Blake the Bible is a model of a text which induces in the reader a transformation, a transubstantiation, a transfiguration from fallen vision to Eternal Vision. But since Eternal Vision includes fallen vision, the Bible contains both perspectives, and so does Blake. Eternal Vision is full of fallen vision. Thus the Bible, like Blake, can be read by some readers as a code book of morality instead of as a means of expanding vision. The reader must actively, creatively, and responsibly read the Bible, just as he must live his life, accountable for his moral stance. Blake does to the Bible what eternal perception does to the fallen world: reveal, oppose, forgive, and transfigure it, reversing it and including it in total Vision.

The transfiguration that Blake works on the Bible is based on a method of presenting oppositions and then transforming them, much as happens at the transfiguration of Jesus (see following chapter for discussion of the transfiguration scene in *Jerusalem*). That is, the Bible clearly stands inside *Jerusalem*; it is essential to the poem. At the same time *Jerusalem* can be said to alter the Bible so radically that it virtually dispenses with the Bible altogether; it surpasses the Bible. Similarly, when Jesus stands transfigured on the mountain, his human body exists within his divine one: the transfigured form clearly includes the earthly form. At the same time the presence of Jesus is altered so radically that his physical form seems unnecessary, completely transcended. From a different perspective, the transfigured form, whether Jesus or *Jerusalem*, reveals the glowing presence that was co-existent with the apparently ordinary form, whether the man of Nazareth or the Bible, all along.

And Blake refuses to compromise this radical combination of divine and ordinary existence. The eternal Vision that is celebrated in the transfiguration scene at the end of *Jerusalem* can be achieved only through an embrace of the fallen world as brought about by the birth of Jesus. Unity and individuality, God and man, transcendence and immanence, minute particulars and Eternity, none of these pairs of supposed opposites can be understood by choosing between them. And neither can fallen vision and eternal Vision, for to choose between them is to fall again, but to see them both is to enter Eternity.

My search for the principle of transfiguration in Blake receives confirmation from David Wagenknecht's idea that a principle of transformation may be the key to Blake:

> As intensive work on Blake continues, it becomes increasingly evident how central and common to all approaches is the idea of transformation. On this common ground meet ways of reading Blake as different from each other (though not necessarily opposed) as Kathleen Raine's and David Erdman's.... Whether or not we want to accept a given read-

> ing ought to give way eventually to a concern for the principle of transformation itself.[10]

Quoting Richard Cody's *The Landscape of the Mind*[11], Wagenknecht sees the idea of pastoral as a compromise between transcendence and immanence. But he intensifies the relationship:

> ... the more apocalyptic the outlook, the greater is man's awareness of his fallen condition. The closer Blake comes to the achievement of imaginative transcendence, the more man comes to seem immersed in a satanic immanence.... Blake's secular and religious concerns are one: to demonstrate that the ordinary world of extensive, fallen vision includes the imaginative wherewithal for that world's intensive, visionary transformation.[12]

Wagenknecht concludes with a reading of the end of *Jerusalem* which agrees with mine: "the paradoxical intercourse between universal and individual ... is the final ... transformation."[13]

Herbert Schneidau, in *Sacred Discontent*, although he mentions Blake only twice in passing, sees the whole development of Western culture as based on a tension between acceptance and rejection. The essence of biblical Hebrew culture and of the Christian and secular cultures which have descended from it, according to Schneidau, is the struggle between continuity and revolution: "We love and hate our culture, and the resultant force is toward change. This ambivalence derives from the Bible."[14] The Yahweh of Israel, even though He is the very foundation of the integrity and continuity of the community, intercedes again and again to discredit the culture of His people and to redefine His relationship to them. Each intervention is simultaneously a destruction of established structures and a construction of new ones.

If Schneidau's thesis about biblical culture and its descendants is accepted, then no one stands more clearly than Blake in the main line of Western cultural development, for Blake insists on redefining both his culture and its Bible even while he claims to honor them. Schneidau traces the dilemma of continuity-in-revolution through Christianity and into modern literature:

> The dilemma is an old one in the Judaeo-Christian tradition: whether to regard the event primarily as that which founds and centers new structures or as that which broke

> away so radically from former structures as to put in question all possible new ones.[15]

This dilemma can lead to a constant surging forward, always hoping, never accomplishing, every structure being undercut. That kind of movement is apparent in Blake; never does he allow the reader to rest content with any oversimplified structural formula. And yet the conclusion of *Jerusalem* is an undeniable triumph, an absolute end. These two contradictory aspects of Blake's poetry create a problem for the reader: if we emphasize the constant revolution too strongly, a vicious cycle results; if we emphasize the construction of systems too strongly, dogma threatens. We have seen the vicious cycle in "The Mental Traveller"; we have seen dogma in *An Essay on Man.* We have seen the near impossibility of escape in Coleridge, Mary Shelley, and Percy Shelley. We have seen the promise of a solution in Nicholas of Cusa. And we see the triumphant solution in Blake.

The eschatology of the early Church, as explained by Rudolf Bultmann in *Theology of the New Testament*, provides a very Blakean perspective on the problem. In Bultmann, an absolute end is paradoxically combined with a hope for the future. Because the Christian feels the urgency of the meaning of Jesus so strongly, time is effectively ended:

> The consciousness that man's relation toward God decides his fate and that the hour of decision is of limited duration clothes itself in the consciousness that the hour of decision is here for the world, too.[16]

The paradox of the kingdom of God is that it is "future and yet already present." The individual believer is torn out of ordinary historical time by a de-historicized and de-sacralized God and forced to confront his true history, his de-secularization, in concrete encounter with his neighbor. I think that it is important not to sacrifice either the individual or the universal aspect of Blake's eschatology. Ronald Grimes attempts to distinguish Bultmann's view from Blake's:

> Blake's eschatology is a matter of renewed vision, but the consummation is not complete until the new personal consciousness has become a new social and cosmic conscious-

> ness.... Bultmann's existentialist eschatology remains on a personal and subjective level.[17]

Similarly, Thomas Altizer attempts to separate the individual and the universal:

> Orthodox Christianity ... has proclaimed an individual redemption that takes place without affecting the reality of the world; but radical Christianity refuses a redemption which is confined to individual selfhood, and seeks an apocalyptic transformation of the world.[18]

Whether or not Bultmann makes that distinction, Blake does not. In Blake there is no separation of the individual's personal transfiguration and the transfiguration of the cosmos.

In Bultmann, the paradox in time extends to include an ethical one: God's reign is not a demand for good. "It aims neither at the formation of 'character' nor at the molding of human society." The fulfillment of God's will is "nothing else but true readiness for it, genuine and earnest desire for it."[19]

A similar affront to ordinary understanding is proposed by Stanley Fish, in *Self-Consuming Artifacts*, as the distinguishing characteristic of a dialectical, rather than a rhetorical, literary method. Rhetorical literature satisfies the reader, telling him what he already knows (as in Pope's "What oft was thought but ne'er so well express'd"), whereas dialectical literature disturbs, often humiliates, acting as a "good physician" who urges a conversion. The dialectic demands a radical new life from the reader: to remain unchanged is to fail to understand.

In his discussion of the *Phaedrus* Fish sees a method in which each part of the work invalidates the part before it. The contradictions and non-sequiturs force a larger perspective of understanding. The technique is not based on logic and reason: "what is being processed in the *Phaedrus* is not an argument or a proposition, but a vision."[20] The reader of Blake must enter into that same spirit of loving confrontation to create his own reconstruction of *Jerusalem*, which is Blake's reconstruction of the Bible.

Most writing, whatever the intention of its author, can be read, if the reader insists, as a moral lesson, a rhetorical confirmation of principles which we already know, and which, if we apply them gradually to our lives, will improve us.[21] The oral teachings of a master dialectician, such as Jesus or Socrates, however, engage the students and disciples in a direct and surprising way, shattering complacencies, preventing that easy and self-satisfied kind of learning. Even though all that we know of Jesus and Socrates has come down to us through writing, it is significant that neither of them was a writer. In fact writing itself necessarily distorts their teachings.[22] As soon as the teachings of a master, especially a master dialectician, are written, they begin to be codified and to lose some of their power. This process occurs not only because the writers may intentionally alter the teachings, but because of the very nature of writing itself. Whatever advantages the medium of writing may have over a spoken dialectic—longevity, logical progression, linear sequentiality—it also has disadvantages. It does not allow for genuine dialogue; it does not allow for certain modes of simultaneity. Blake created a form of art which attempts to transcend and revolutionize methods of writing, even while it employs them. Logic and sequence are blasted by paratactical strategies; contradictions and paradoxes halt the reader; illustrations violate or ignore the text. Such a poet is quite simply incomprehensible to anyone who sits down to read complacently. In effect Blake found a way to confront his reader in his text as Jesus and Socrates confronted their listeners in person.

The reader does not so much have to go away and contemplate the text in tranquility, as he has to confront the Eternity of Blake's minute particulars with every sweep of the eyes across the words and illuminations. And such a confrontation is contained right at the beginning of *Jerusalem* in the phrase "Monos o Iesous." (See further discussion in Chapter 8.) The coincidence of opposites—the forgiving immanence of the woman taken in adultery and the judging transcendence of the transfiguration—teaches us by embodying, not by merely urging and preaching, the expanded consciousness of contradictions.

The dialectic is not only the cause of the method and the object of the method; it is the method itself. The simultaneous imitation and rejection, the simultaneous immanence and transcendence, of all experience and of the Bible in particular, teach us a new way to read and a new way to live, a way that demands awareness of how our ordinary understanding distorts everything from religion to sex, from perception to philosophy. In Plate 3 Blake follows the example of St. Paul in declaring himself the greatest of sinners; if we deny our own sin, or any undesirable quality that we can abstractly conceive, then we deny ourselves dialogue with Jesus.

Exploring Blake from philosophical perspectives, Leopold Damrosch, in *Symbol and Truth in Blake's Myth,* finds four categories of irreconcilable differences: epistemological (sense/intuition); psychological (solipsism/universal humanity); ontological (divine immanence/transcendence); and aesthetic (imaginative/ fallen art). He claims that Blake cannot come to terms with oppositions because of his refusal to compromise: "Rather than accepting one or another of the compromises that have been developed over the centuries, he aspired to put the entire structure of Western thought together again."[23] But always Damrosch refuses to accept the contradictions which Blake presents; when he finds logical inconsistencies, he stops short: for example, the meanings in "The Blossom" are "mutually contradictory: if the body is a prison, then it is not a source of joy" (112). In a related issue very important in *Jerusalem,* he relegates Blake to the realm of mystical vision which is unintelligible to all non-mystics:

> ... the body is at once a merciful 'limit of contraction' and a trap from which we must escape. But it is easier to say that it is both at once than to understand how it can be.... I do not deny that analogues to Blake's position may be found in Boehme and elsewhere; I deny that they make sense.[24]

Damrosch's book provides an invaluable service by continually confronting the paradoxes in Blake, and many times, despite his refusal to enter fully into Blake's Vision, he formulates concepts in a precise and revelatory way: writing of Emanations, he thinks that they, like many other parts of Blake, "must similarly be understood as a mystical attempt to keep what we have and yet transform it utterly." He realizes that Los's work is important, but that it is not Eden. He concludes that Blake's symbols must point beyond themselves to the truth and that Blake's myth believes in man's spiritual power "while fully recognizing the self-deluding tendencies of the imagination and its symbols."[25] But finally Damrosch stands forlornly outside:

> ... if we inhabit a world that no longer believes in its symbols—if we can neither trust the products of our symbol-making imagination nor bear to live with them—then

> Blake speaks to us with a special poignancy. His Eden is forever closed to us by the Cherub with the flaming sword.[26]

Damrosch's thought is an extreme consequence of approaching Blake with too much of a commitment to logic. Despite his genuine insights into Blake and his obvious affection for Blake, Damrosch's Covering Cherub, which forbids his entrance into Blake's Eden, is his insistence on reasonable, logical, coherent, philosophical systematizing as a way in. Instead, the recognition of the coincidence of opposites can provide the tool for entering Blake's vision.

A rational approach to the Bible sees it as a code of morality, but Blake's Bible is not a code of morality; instead, as declared in his most famous pronouncement on the Bible, in *The Laocoon*:

> The Old and New Testaments are the great Code of Art. (E273)

Peter Fisher, in *The Valley of Vision*, presents a good formulation of the accepted interpretation of this statement:

> Blake called the biblical record "the Great Code of Art" not because it outlined the rules of composition, but because it presented a collection of literary forms inspired by the Hebrew genius who was Jehovah in the Old Testament and Jesus in the New. It had an inevitable pattern undistorted and unrestricted by the accidental events of the narrative.[27]

However, Blake's word "code" contains a contrary tension, like that in the word "system."[28] In every one of the six other times that Blake uses the word "code" or "codes"[29] it is a pejorative term, always referring to a divinely inspired law given for the purpose of restriction or war. The most vehement use occurs in the annotations to Watson:

> The laws of the Jews were (both ceremonial & real) the basest & most oppressive of human codes, & being like all other codes given under pretence of divine command were what Christ pronounced them The Abomination that maketh desolate, i.e. State Religion which is the Source of all Cruelty. (E618)

Any code which pretends to divine inspiration is liable to usurp the place of God. Of course the phrase has the positive meaning that Fisher assigns to it, but at the same time, any code, because it exists in the fallen world, must partake of the nature of that world.

One of Blake's earliest works (generally assumed to be his first attempt at illuminated printing), the tractate "All Religions Are One," helps to clarify what is at stake here:

> The Jewish and Christian Testaments are an original derivation from the Poetic Genius. This is necessary from the confined nature of bodily sensation. (E1)

Of course the common interpretation of this statement, as exemplified by Fisher above, is correct, but working at the same time is the restrictive nature of the Bible and the body. As usual in Blake, if the interpreter too much emphasizes one extreme in Blake, that interpreter is probably missing the equally strong contrary movement. In the same tractate, Blake states: "As all men are alike in outward form, so (and with the same infinite variety) all are alike in the Poetic Genius." Just as each man's body is a particular expression of the human form, so each religion is a particular culture's expression of the Poetic Genius: "The Religions of all Nations are derived from each Nations different reception of the Poetic Genius." Any particular body, or any particular Bible, is an expression of "the Poetic Genius," which "is the true Man" (E1).

But the limited shape of the human body, just like the limited shape of the Bible, just like any code or any system, can be seen in its exclusive sense rather than its universal one. Imaginative acts must take definite shape; if they remain undefined, then man is forever lost in the void. But whatever shape they take is determined by both their eternal and fallen nature. The eternal exhibits itself in the temporal, fallen world; the distinctions in the fallen world derive from eternal unity. The Christian artist must honor the gifts of God in other men and in other bibles, not because each one is a Platonic shadow of something more real, but because each one is eternal if perceived with expanded Vision.

Fallen vision does not perceive the eternal, and so tries to create its own substitute for eternity by concealment and mystery.

Eternal Vision perceives the essential coincidence of opposites which fallen vision falsely divides, and so fully enters into the definite shapes, the minute particulars where the center and the circumference of Eternity meet. But once such a particular shape is entered into, the Fall happens again. Jesus is born in *Jerusalem* and gives himself a definite shape, in order to break through the false categories that fallen vision tries to maintain. Jesus submits to the fallen world in order to reveal it for what it is: the eternal perversely reflected.

When Blake describes the process by which *Jerusalem* was composed, he unexpectedly incorporates the Old Testament God who gave the Ten Commandments on Mount Sinai:

> Reader! lover of books! lover of heaven.
> And of that God from whom all books are given,
> Who in mysterious Sinais awful cave
> To Man the wond'rous art of writing gave,
> Again he speaks in thunder and in fire!
> Thunder of Thought, & flames of fierce desire:
> Even from the depths of Hell his voice I hear,
> Within the unfathomd caverns of my Ear.
> Therefore I print; nor vain my types shall be:
> Heaven, Earth & Hell, henceforth shall live in harmony.
> (*Jerusalem* 3:1–10, E145)

Blake includes here the tradition that God gave man the gift of writing on Sinai.[30] That art can be used to record a restrictive code like the Ten Commandments or it can be used to unite Heaven, Earth, and Hell. Blake insists on both modes of writing: the restrictive writing which comes from a terrifying God in a secret place is also a means of regeneration. The code of prohibitions arises from a shrunken perception of the nature of writing; as a list of negatives ("Thou shalt not"), it negates the visionary power which is visible to expanded perception. But expanded perception does not negate the cave or the fearful God or the limitations of writing. It sees them anew, assimilating while overturning. That which had been denied, the prohibitive Law from Sinai, is now included in a new totality of Vision. Total Vision does not simply exorcise the old punitive God; it assimilates Him.[31] Similarly Vision must use the tools and work with the limits of the fallen world. A Heaven which excludes Hell, a

Hell which excludes Heaven, or an Earth which excludes either one will not enter the Savior's kingdom. All that exists must be revealed, because refusal to acknowledge any part of existence in itself negates the transforming power of Vision.

But this inclusiveness necessarily entails redefinitions, of both the new covenant and the old. The exterior, corporeal thunders and fire have been interiorized. In one sense, this makes them less fierce: they are contained. In another sense, it makes them all the more terrible. Now instead of an unfathomable distant heaven, the poet feels the terrors of the unfathomable inside his own mind and body, in an exhilarating and terrifying coincidence of opposites.

Blake strongly insists on both aspects of his God, the transcendent and the immanent:

> We who dwell on Earth can do nothing of ourselves, everything is conducted by Spirits, no less than Digestion or Sleep.... When this Verse was first dictated to me I consider'd a Monotonous Cadence like that used by Milton & Shakespeare.... But I soon found that in the mouth of a true Orator such monotony was not only awkward, but as much a bondage as rhyme itself. I therefore have produced a variety in every line, both of cadences & number of syllables. (E145–46)[32]

These statements contain a clear and uncompromising paradox: on the one hand, the poet has no power of his own; on the other hand, the poet decides how to write his poetry. The assertions about the process of composition contradict the theory of all-powerful spirits; the word "consider'd" contradicts the word "dictated" just a few spaces before it. Some critics find Blake simply confused here; most ignore the transcendence and make Blake's God completely immanent. But surely Blake wants it both ways: not only does he use the terminology of a transcendent, all-powerful God, but he carefully sets up the tone of awe which such a God induces. The very art of writing which Blake is using to escape from mystery and caves is given from a "mysterious" cave. Just as any code or system necessarily participates in the fallen world, so language itself cannot, and should not try to, totally escape its fallen nature.

> For Blake, the claim of individual expressive authority and the disclaimer of authority ... involves no contradiction, for the universal poetic genius that is God acts only through individuals. That is why Blake can seem to be both the author of original writings and merely a conduit through which innumerable writings ... transmit themselves.[33]

Imitating his God even while he rebels against Him, fearing His transcendent power while internalizing Him, the poet performs his paradoxical task.

> Again he [God] speaks in thunder and in fire!
> Thunder of Thought, & flames of fierce desire:
> Even from the depths of Hell his voice I hear,
> Within the unfathomd caverns of my Ear.
> Therefore I print; nor vain my types shall be:
> Heaven, Earth & Hell, henceforth shall live in harmony.
> (*Jerusalem* 3:5–10, E145)

With his only use of the word "types"[34] Blake includes not only a direct reference to his techniques of printmaking, but also a hint of his method of typology and the kinds of reversal and fulfillment contained within it. Blake used very laborious and exact engraving and etching methods to produce his illuminated books, as well as his commissioned works to earn his livelihood.[35] He first wrote and drew on copper plates with a wax ground. A corrosive acid then burned away the exposed surfaces, leaving only the design to be printed. The copper plate was inked and pressed onto paper. The raised surfaces on copper are literally the type of the finished product on paper. But the entire physical process of printing is figuratively a type of the spiritual process of regeneration. Although it has its own identity as a physical process, it fulfills itself only in its spiritual or mental final product.

The printmaker can complete his task only in an action of physical reversal. That is, when he prints on paper, his design and his words are reversed.[36] Throughout *Jerusalem* Blake indicates that the attainment of eternal Vision can come only through a reversal of the fallen world.[37]

In a profound sense, to find the eternal world is to reverse the fallen world, just as the printmaker must reverse his design in order to print it onto paper. But he cannot reverse the plate until he has fully shaped it, in all its minute particularity. Furthermore, the design on paper is identical in every detail to the design on copper, except that it has been completely reversed, transformed in its perspective as well as in its medium. It is entirely different, even while it is the same, and both the sameness and the difference have been radically redefined from their original connotations. The physical importance of the simultaneous identity and difference strikes us immediately when we encounter the mirror writing in *Jerusalem* on plates 37 and 81, and yet that process of transferring mirror images is embodied in every single plate of Blake's illuminated works. Its spiritual importance, similar to its physical importance and yet quite different, also strikes us when we encounter on almost every plate the stunning redefinitions and rewritings that are the soul and method of *Jerusalem*.

Just as the printmaker's work must be fulfilled by a process of reversal, so the Christian artist's task must also be fulfilled by a process of reversal, a reversal which completely accepts the fallen world and at the same time utterly transforms it. The dialectic is like that of the story of the woman taken in adultery. Jesus forgives the sin by simultaneously accepting it (refusing to punish, implicating the accusers) and rejecting it (not condoning it, telling the woman to "go and sin no more").[38] This paradoxical behavior cuts across old categories and, if not ignored or avoided or explained away, re-creates the world in a new way. Likewise, the Bible, most clearly in the relationship of the New Testament to the Old (especially in the entrance of Jesus into the world) contains a paradoxical conjunction of fulfillment and reversal. Although Jesus does come to fulfill the old law, he also comes to destroy it and replace it with a new law. And Blake's dialectical relationship to the Bible performs a similar work on it. Accepting the Bible as his model, both explicitly by declaration and implicitly by quotation and imitation, he nevertheless burns away its falsehoods as the acid burns away apparent surfaces, and reverses its perspectives by re-writing it, remaining totally faithful to it while utterly transfiguring it.

Christine Gallant points out that in the early Lambeth books (*Book of Urizen, Book of Ahania, Book of Los*) Blake finds himself in a paradox when he tries to fight the rigidities of myth by constructing his own myth, which is in danger of becoming too rigid itself. There is a necessity for clear outlines, according to

Blake's aesthetic, political, and religious beliefs, but clarity can become a Urizenic mistake.[39] To put it in its bluntest form, Urizen's earliest impulses toward fixed form are Blake's own.[40] Gallant traces Blake's grappling with this problem, especially through *The Four Zoas*, and shows that by the end of that poem a "new imaginative unified vision" includes in Golgonooza that which it had originally tried to transcend. Using Jungian terminology, she writes that the incomplete mandala has been completed from the centers of energy in the psyche:

> Christ has thus truly acted as an archetype of the Self for Los ... showing him by example how to construct a 'city of art,' which, paradoxically, will have as its essential ingredients all that it originally had been built to repulse.[41]

Michael Cooke takes an even more extreme view of Blake's inclusiveness: the movement is not towards action but rather towards a "complex condition of spirit." According to him there is no final resolution: "The crucial factor is a matter of mode or mood of vision, or what one makes ontologically of oneself and one's situation."[42] Karl Kroeber redefines the conflict in *Jerusalem* by writing that there is no division of the sacred and the profane in the poem. Every atom is equally sacred, and the conflict is between those who incorrectly make the division between sacred and profane on the one hand and the power of Jesus on the other, which unveils the intrinsic divinity of every minute particular.[43]

All these critics are working toward a vision of *Jerusalem* which acknowledges the frightful conflicts that dominate much of the poem but which also acknowledges the final harmony that is established. Those critics who emphasize the irreconcilability of conflict imply, if they do not state it directly, that in the final resolution the forces of right (Blake's side) conquer the forces of wrong (the opponents of Blake). Those critics who emphasize inclusiveness imply that concepts of right and wrong must be redefined. Both critical positions find support in the text because both positions are there. Blake's apocalypse is obviously a triumph that wins over enemies, but it does not exclude the supposed losers. "The Glory of Christianity is, To Conquer by Forgiveness" (*Jerusalem* 52). The principle of opposition, which according to Frye structures each chapter,[44] is indeed there, but is itself opposed by a principle of inclusiveness. The two principles appear to be mutually exclusive, and yet both are clearly there. Our reason is confounded by the coincidence of opposites and forced to seek a larger vision.

On the way to this larger vision, Los insists that it is important not to make the wrong distinctions and absolutely essential to make the right ones. On Plate 7 he contrasts the regeneration of the eternal resurrection to the temporal generation of a vegetated Christ:

> In anguish of regeneration! in terrors of self annihilation:
> Pity must join together those whom wrath has torn in sunder,
> And the Religion of Generation which was meant for the
>   destruction
> Of Jerusalem, become her covering, till the time of the End.
> O holy Generation! Image of regeneration!
> O point of mutual forgiveness between Enemies!
> Birthplace of the Lamb of God incomprehensible!
> (7:61–67)

But most important is the dispute over categories commonly accepted by Christianity:

> And this is the manner of the Sons of Albion in their
>   strength
> They take the Two Contraries which are called Qualities,
>   with which
> Every Substance is clothed, they name them Good & Evil
> From them they make an Abstract, which is a Negation
> Not only of the Substance from which it is derived
> A murderer of its own Body: but also a murderer
> Of every Divine Member: it is the Reasoning Power
> An Abstract objecting power, that Negatives everything
> This is the Spectre of Man: the Holy Reasoning Power
> And in its Holiness is closed the Abomination of Desolation
> Therefore Los stands in London building Golgonooza.
> (10:7–17)

The two contraries, which are inextricably part of every substance, are abstracted and separated. Without the false separations (the lies which abstract reason promulgates) good and evil could never be taken apart in the first place. In fact such a separation negates the very essence of life itself. Traditional

Christianity (not to mention many a Blake critic) makes a distinction between good and evil and assigns the former to the sheep and the latter to the goats at the Last Judgment. So the traditional Last Judgment is a form of mass murder, whose purpose is to reassure those who abstract and separate in their exclusive self-righteousness. The self-righteous false holiness that underlies such a system objects to every definite act as a sin (*Jerusalem* 80:53) and in its incarnation as the Spectre, discourages the poet from acts of clarification and forgiveness.

Los specifically counters that system in building Golgonooza: "Therefore Los stands in London Building Golgonooza" (10:17). If Los does not build, then he becomes a victim of the system of abstract religion with its mystifications. He does not construct his new system in order to define a new tyranny, a rock-built refuge from which he is unassailable, but in order to prevent something worse from happening. He seems to believe that if he can just take actions exactly opposite to those of the Sons of Albion, then he can combat them and solve the problem:

> I must Create a System, or be enslav'd by another Mans
> I will not Reason & Compare: my business is to Create.
> (10:20–21)

In this much-quoted line from Blake, too many critics simply read Blake's desire to create his own system. However, Blake is not only opposing a specific system, but all methods of systematizing which deny creativity. As is the case with the word "code," the word "system" is double-edged. In the few other places where Blake uses the word "system," it is clearly pejorative.[45] The important impulse here is creation, not system building. To build a coherent and consistent system, the poet would have to reason and compare constantly, but that is exactly what he will not do. He creates in order to clarify, but not in order to systematize. It is crucial that Los does not get trapped in the same kind of abstractions of good and evil, deriving from abstract reasoning, that he is combatting in the Sons of Albion. In fact, in Chapter 2 he will discover that even the admirable distinctions which he wishes to make are not easily made against his own systems. That is, Los's system creating has as its purpose the deliverance of all individuals, including himself, from all systems, including his own. His aim is to destroy slavery, not impose a new tyranny.

One critic who tries to capture Blake's refusal to construct a coherent system is Michael Cooke:

> Blake ... seeks but to escape some other person's prediction; his 'system' is to be formulated, as a sort of perpetually indefinite defensive maneuver.[46]

Blake's system making does have the dynamic quality suggested here, but Blake does not remain indefinite. As in Bultmann's eschatology (see above), Blake's apocalyptic end is both present now and still to come. It is in fact this unique and complex vision incorporating both a refusal to be bound by any system and at the same time a refusal to remain indefinite which is the essence of Blake and of his use of the Bible. The activity of reasoning and comparing may be the work of the philosopher, the theologian, and the critic, but not of the poet of the imagination. Instead of serving life-denying abstractions, what Blake calls "Negations," the words of the poet must use language in the service of some higher and deeper reality which denies and at the same time affirms the power of the words.

The major difference between Los's work and that of the Sons of Albion is in revelation and definition, not as abstractions but as complete and carefully outlined actions. The Sons make false distinctions; good and evil cannot be separated, as Los knows. Los's, and Blake's, work gives shape to truth and error, good and evil, pleasure and pain, and does not try to deny or conceal either apparent side of an opposition. The fight is against denial, concealment, and doubt. All that exists must be revealed (the basic meaning of the word "apocalypse") so that error can take on its clearest and most powerful shape, in particulars and in the aggregate, and finally fall away under its own dead weight, snared and taken by its own lying power, reversed and incorporated by the presence of Jesus. Even more than in Shelley's myth, total revelation brings reversal of tyranny.

The work of the Sons of Albion tries to consolidate the reasoning power, not to reveal it, but to hide it and maintain its negating force. However, the splendid irony of *Jerusalem* is that even this work which sets itself against revelation, even this work which attempts to solidify and enshrine an abominable holiness in its center must eventually be converted into part of the greater unity. As long as the holy secretiveness at the center—whether that of the original tabernacle and Temple or that of the usurping Abomination of Desolation which is the same force to a higher power—tries to maintain itself, it is caught in the tomb of death-in-life. But even this tomb reveals itself to be also the site of the resurrection, life-out-of-death. Its force consolidates itself until it must reveal the self-destructive negation which reverses it. Once all is revealed in the resurrection/apocalypse/transfiguration, then doubt must disappear in its present form and all contraries be incorporated into the whole.

The difficulty in reading Blake is not only that his definitions do not always seem clear; it is also that they often appear self-contradictory. Both of these confusions occur because Blake affronts our common sense with his uncommon Vision. The theology of the Sons of Albion is surely consistent, coherent, and rational, because it is formulated from abstractions around a holy center. Blake's theology, however, is inconsistent and anti-rational because he is pursuing the details of the world, full of life and therefore of oppositions. He creates instead of comparing, and finds his holiness in the circumference, which is identical to the center, where humans meet each other and meet Eternity:

> What is Above is Within, for every-thing in Eternity is
> translucent:
> The Circumference is Within: Without, is formed the Selfish
> Center
> And the Circumference still expands going forward to
> Eternity.
> And the Center has Eternal States! (71:6–9)

If truth and error, or good and evil, were easily distinguishable, then Los could abstract his principles as the Sons of Albion do, make a few general rules of morality, and build his system in relative ease and certainty. And indeed, many critics of Blake see him doing just that. But the agony of Los, and of Blake, in *Milton*, *The Four Zoas*, and *Jerusalem*, is the need to confront complexity. The certainty of faith is in revelation, not in any abstract formula, in concrete encounters, not in general rules. Because this process of revelation takes place in every word in *Jerusalem*, in effect the apocalypse is taking place in every word. If the truth were easy, then Los would not need to suffer all the tears of building Golgonooza. But it is painful to destroy what one is working for as part of the process, especially to undergo the annihilation of Self.

The prophet-poet must rebel against any secretive establishment that imposes abstract morality, even if it means making his own mistakes, not to set up a rival regime, but to force error into revelation of itself. Los does not build Golgonooza to establish a permanent monument, but to define the problems more clearly. The problem can never be defined, and therefore never solved, if so-called good and evil are separated by fiat, good locked safely away untouchable, and evil banished from acknowledgement. Such a scheme denies the inextricability of the qualities with which every substance is clothed, and thereby murders its own body and lays it in the tomb like Lazarus or the crucified Christ. Such a system must be broken by the power of creativity which allows the birth of Jesus through forgiveness of sin and sees the transfiguration in all its minute particulars.

The Sons of Albion are so afraid that they will lose their identities that they cannot allow their Selfhoods to be broken down. The irony is of course that if they are not broken down voluntarily, they will break under their own self-destructive impetus. Error will be reversed against its will, unless it is willing to annihilate itself in the furnaces, as Albion does on Plate 96. The Sons fear error and evil and take strenuous measures to protect against them; at bottom they do not trust any scheme which is not abstractly, coherently moral; they need to protect themselves from the encroachment of the rich details of life. Blake sees most of religion and philosophy enslaving themselves to this system of abstraction, and thereby assuring the entombment of the very life they seek and love while enshrining the Abomination which desolates humanity's hope. The work of Los accepts the necessity of living in error and evil, but is based on the fundamental optimism that ultimate revelation will lead to the salvation of us all. Trying to be abstractly consistent works against the very goals which a religion desires: the finding of the eternal in the temporal. Only by losing the self can one find it: "whosoever will lose his life for my sake shall find it" (Matthew 16:25).

While waiting for Jesus to rend the veil, Los, in order to prevent Albion from turning his back against the Divine Vision, descends into "the interiors of Albions/Bosom, in all the terrors of friendship" to "search the tempters out" (43:3–5). Erdman calls this "a Diogenes-like search ... which is the central action of the whole poem ... and [which] is shown beginning in the frontispiece."[47] But Los finds the task of destroying the punishers and sparing the victims completely impossible:

> [Los] saw every Minute Particular of Albion degraded & murdered
> But saw not by whom; they were hidden within in the minute particulars
> Of which they had possessed themselves;
> What shall I do! what could I do, if I could find these criminals
> I could not dare to take vengeance; for all things are so constructed
> And builded by the Divine hand, that the sinner shall always escape,
> And he who takes vengeance alone is the criminal of Providence;
> If I should dare to lay my finger on a grain of sand
> In way of vengeance; I punish the already punished.
> (45:7–9, 29–34)

William Butler Yeats emphasizes the passivity which is necessary at this point: "the tomb of Christ could be no other than a shelter, where imagination might sleep in peace until the hour of God should awaken it."[48] Although David Wagenknecht slides around the issue, he hints at the double nature of the couch:

> The period of the clarification of error, while a place is being prepared for Satan and while man has simply in his divided being to wait, is given in chapter 2 an ironic identification and structure: it is the Scriptures ..., organized around the sixteen books of the Old and New Testaments most important to Blake.[49]

In their concentration on the constructive activity of Los, too many critics overlook the wise passivity perceived by Yeats and Wagenknecht; both the passivity and the activity must be acknowledged. It is necessary to wait for the coming of Jesus, and it is just as necessary to continue to work while waiting, as Los does.

It is absolutely essential that Blake's *Jerusalem* itself, as a re-creation of the Bible, be seen in this double way. On the one hand Blake the poet is hammering out his own destiny, refusing any compromise. On the other hand, he is completely dependent on the divine power which is gracefully vouchsafed to him. This paradox is expressed right from the beginning of the poem when the poet apparently contradicts himself about his method of composition (see comments on Plate 3, above). The apparent paradox arises because ordinary language is inadequate to describe the Vision necessary to perceive the relationship between the fallen and the eternal. Similarly, the couch of Albion is both terror and mercy, both fallen and eternal. In the effaced words on Plate 1, Albion's Couch, England, is both a globe in the void and a pleasant shadow of repose. Thus, in the various connections which Blake draws, the Bible is analogous to or identical to a series of sculptures and to a couch which in turn is equated with a void, a womb, a tomb, and with England, which has been identified with Canaan. All of these symbols have a double meaning, but the doubleness does not consist merely in seeing the same thing in two different ways. They must be seen as simultaneously fallen and eternal, as sinful and at the same time necessary for redemption, accepted and rejected together.

In a simple optical illusion, for example the famous one where the observer can see either two profiles or a vase, it is possible to see both things at the same time with the proper mixture of concentration and relaxation. Stephen Prickett disqualifies this kind of vision when he reproduces this illustration in his *Words and the Word*. He cites E.H. Gombrich, who

> reminds us in connection with figures like this that an ambiguity, as such, cannot be perceived. 'To perceive,' means, in this context, to form a complete and reasonable plausible image—even at the risk of excluding other plausibe, but contradictory images. We can never 'see' directly something as ambiguous; we can only infer that it is so by a process of making first one reading and then another until all possible configurations are satisfied. In this case we can see *either* a vase, *or* two faces in silhouette; we cannot, however hard we try by switching rapidly from one interpretation to the other, 'see' both at the same time.[50]

This stand recalls Barfield's emphasis on Coleridge's polarity, a dynamic back-and-forth between irreconcilable oppositions.

However, I insist that an observer can see both the vase and the faces at the same time; I know, because I can do it. Although I have not seen the face of God, nor communicated daily with my dead brother Robert, nor written *Jerusalem*, I can enter Blake's vision to the extent of living in the coincidence of opposites in that particular optical illusion. This Blakean vision is not simply "ambiguity" or "polarity." It is rather a dynamic coming together of irreconcilable elements, which we can achieve if we persist in our folly. If the reader can bring to *Jerusalem*'s paradoxes the same inextricable mixture of activity and passivity which the poet brings, then she can perceive things which she thought were mutually exclusive, and she will be seeing no illusion, but Vision.

The distinction between Negation and Contrary is crucial in Blake. Damrosch, for example, obscures the difference by claiming that Blake defines out of existence the parts of life which he does not like.[51] But Negations are not just the aspects which Blake does not like; they are the refusal to allow life to exist at all. A useful analogy can be made with the physical process of etching and printing that Blake employed. Negations are analogous to the surfaces which must be burnt away by the corrosive acid, the doubts and despairs of the Selfhood which must be burnt away by the fires of the Last Judgment, the veil which blocks Vision.[52] Contraries are analogous to the surfaces which are protected by the wax ground (the merciful protection of the limits, the couch of Albion, the Mundane Shell), and then reversed onto paper to be fulfilled. If the Negations are not burnt away, then the printed design will be blurred and indefinite. The more sharply the designed surfaces are engraved and etched, the more exact will be the minute particulars finally printed.

The Spectre in *Jerusalem* does not understand the difference; he thinks that he is a Contrary:

> ... the Almighty hath made me his Contrary
> To be all evil, all reversed & forever dead: knowing
> And seeing life, yet living not.
>
> (10:56–58)

Los, however, who is busy building systems to deliver from systems, insists on the distinction:

> Negations are not Contraries: Contraries mutually Exist:
> But Negations Exist Not: Exceptions & Objections & Unbeliefs

Exist not: nor shall they ever be Organized for ever & ever:
If thou separate from me, thou are a Negation: a meer
Reasoning & Derogation from me, an Objecting & cruel Spite
And Malice & Envy: but my Emanation, Alas! will become My
Contrary:...
... never! never! shalt thou be Organized
But as a distorted & reversed Reflexion in the Darkness
And in the Non Entity ...
...
And if any enter into thee, thou shalt be an Unquenchable
Fire
And he shall be a never dying Worm, mutually tormented by
Those that thou tormentest, a Hell & Despair for ever & ever.
(17:33–47)

To make it clear that the Spectre is the recalcitrant part of Los himself, Blake immediately explains, "So Los in secret with himself communed" (17:48). Thus the denial of the Spectre and the refusal to organize him is not a contradiction of Los's method of revelation, but is a denial of denial itself. Not only are the doubt and despair of the Spectre a Negation of all life and action, his very power of abstract reasoning is a Negation. It is that power of abstract reasoning which led the Sons of Albion to construct the Abomination of Desolation, and which leads the Deists to accept a reasonable and rational view of human life which denies the need for Jesus, which in fact perceives Jesus as an illusion. Accepting nature, and human nature, as they are apparently given, the Deists fail to burn away the apparent surfaces to achieve the finished design, and thereby plunge themselves into a hell of their own making.

Blake's simultaneous fulfillment and rejection of the Bible is epitomized in his rewriting of the birth of Jesus. His version is based on that of the gospel of Matthew:

> Now the birth of Jesus Christ was on this wise: When as his mother Mary was espoused to Joseph, before they came together, she was found with child of the Holy Ghost. Then Joseph her husband, being a just man, and not willing to make her a public example, was minded to put her away privily. But while he thought on these things, behold, the angel of the Lord appeared unto him in a dream, saying,

> Joseph, thou son of David, fear not to take unto thee Mary thy wife: for that which is conceived in her is of the Holy Ghost. And she shall bring forth a son, and thou shalt call his name Jesus, for he shall save his people from their sins. Now all this was done, that it might be fulfilled which was spoken of the Lord by the prophet, saying, Behold, a virgin shall be with child, and shall bring forth a son, and they shall call his name Emmanuel, which being interpreted is, God with us. Then Joseph being raised from sleep did as the angel of the Lord had bidden him, and took unto him his wife: And knew her not till she had brought forth her firstborn son: and he called his name Jesus. (Matthew 1:18–25)

Blake's version picks up the story *in medias res,* while Joseph is thinking on Mary's pregnancy. Although Matthew's account does not tell us that Joseph spoke to Mary about his pending decision, Blake thrusts his reader into the heat of the imagined dispute. In effect he externalizes the internal thoughts of Joseph and has Mary participate:

> [Jerusalem] looked & saw Joseph the Carpenter in Nazareth & Mary
> His espoused Wife. And Mary said, If thou put me away from thee
> Dost thou not murder me? (61:3–5)

This Mary immediately puts the lie to Joseph's attempt to be a "just man." In the self-righteousness of his private counsel, he thinks that it is better to have a quiet divorce than to subject Mary to public ignominy. In fact the Old Testament law allowed for a simple end of a marriage:

> When a man hath taken a wife, and married her, and it come to pass that she find no favour in his eyes, because he hath found some uncleanness in her: then let him write her a bill of divorcement, and give it in her hand, and send her out of his house. (Deuteronomy 24:1)

Mary, however, appears to be remembering an even harsher punishment:

> And the man that committeth adultery with another man's wife, even he that committeth adultery with his neigh-

> bour's wife, the adulterer and the adulteress shall surely be put to death. (Leviticus 20:10)

From Blake's point of view, the self-righteous putting away of the sexually active woman is just as bad as murder. In the story of the woman taken in adultery, death by stoning is exactly what the accusers desire for her. By denying the life-giving force of sexuality, Joseph, trapped in the punishments of the old law, would in effect kill his wife-to-be. His attitude is caused by and would cause the absence of Jesus, death without resurrection. Up to this point in *Jerusalem* the birth of Jesus has been looked forward to as a reversal of disaster and a deliverance from mental slavery. The process of Generation which is evidence of fallen sexuality must be allowed to continue so that Jesus can be born.

If Generation is denied, either by the abstractions of the Sons of Albion or by the self-righteousness of Joseph, then its transformation into Regeneration is also denied. If the man rejects and attempts to punish the sexual activity of the woman as sin, then he must at the same time reject his own sexuality as sin. (It is, in effect, this confrontation of the implications of accusing others of sin that Jesus uses to turn away the crowd of potential rock-throwers from the woman taken in adultery; see Chapter 8 for further discussion.) Joseph's denial, if allowed to prevent sexuality, birth, and Generation, would in effect deny the embrace of the fallen world which is necessary for its redemption.

Mary's drawing out of the implications of Joseph's position at first has no salutary effect on him. He is still caught in the language of accusation:

> ... Joseph spoke in anger & fury. Should I marry a
> Harlot & an Adulteress? (61:5–6)

The answer to this rhetorical question, which obviously should be no, turns out to be yes, because, as Jesus clearly expresses, and as Joseph learns to see in this episode, and as Jerusalem gropes toward seeing, so-called sexual sin can be perceived in a completely different way. Joseph is trying to follow rigid definitions which fix human beings into the states through which they can pass, just as the male and female in "The Mental Traveller" do to each other. As is pointed out later in the plate, "Every Harlot was once a Virgin" (61:52). In Blake's theology, it is not necessary that the holiness of Jesus be based on the virginity of his mother.

In fact holiness cannot be based on virginity, for such an orthodox theology creates a Jesus out of self-righteousness and prohibition, that is, abstraction and denial, a Jesus who could never be the Savior of and the friend of sinners. Such a Jesus would be the punitive God of Vengeance. In fact holiness is not the issue.

The birth of Jesus is the ultimate symbol of falling into sin, but it is also paradoxically the triumph of human creativity.

> By his Maternal Birth he is that Evil-One
> And his Maternal Humanity must be put off Eternally
> Lest the Sexual Generation swallow up Regeneration
> Come Lord Jesus take on thee the Satanic Body of Holiness.
> (90:35–38)

Jesus is born into the state of sin so that He can deliver man from accusations of sin. If He is born sinless, of a virgin, then He is not human, is not even God, and cannot deliver anyone. Those who believe in such a sinless Jesus self-righteously project their own sinful feelings onto others, as do the accusers of the woman taken in adultery, and thereby kill the possibility of new life, of Regeneration out of Generation, trapping themselves and others into an unending round of accusation, punishment, and death, as do the characters in "The Mental Traveller." Because they can see only the horrifying aspect of birth, they choose death over life. Therefore, when Jesus wants to comfort Jerusalem, who is being accused of sexual sin and being threatened with punishment, He shows her a scene of Mary in a similar situation. But miraculously Jesus Himself is born from that situation; that is, where the powers of accusation and sacrifice are strongest, there Jesus can appear. By using language from the resurrection of Lazarus just before he shows her the scene, he equates his birth with resurrection.

In the Bible the excuse for redefining the supposed sin of Mary is that it is not really a sin at all. It did not happen. The apparent harlot is a virgin all along. The giant abstract penis of God's Holy Ghost has passed secretly into her and deposited its abstract sperm without violating her holy virginity. The supposed sin has not taken place, and so there is nothing to understand and forgive, simply a mastery and a mystery to accept and submit to. But in Blake the sin is not wished away; instead the event is accepted and redefined in a different way. It is understood instead of being mystified. The apparent harlot is passing through a state of sin, as all humans must do. Blake's method simultaneously deepens the sense of sin and lessens it. The sin in *Jerusalem* is more serious than that in the Bible: it cannot be explained away by recourse to a higher power's authority. But it can be accepted as a fact and then forgiven by the power of a higher impulse in man than the impulse to separate and accuse and punish. Without the forgiveness of this higher spirit, which is Jesus in man, man would be condemned either to repeat the same dull round over and over again or to fall into amorphous oblivion.

Blake's whole Christian theology is founded on the forgiveness of sins. Margaret Bottrall points out that Christianity does not have to be based on forgiveness:

> Blake's exaltation of Forgiveness as the essential quality of the religion of Jesus may seem arbitrary, even to those who are reasonably familiar with their New Testament; for neither in the gospels nor the epistles is there an explicit reiteration of the theme, such as we find in all the later writings of Blake himself.[53]

And Altizer remarks[54] that the New Testament contains no statement of forgiveness as Blakean as that in Jeremiah 31:34:

> … they shall all know me, from the least of them unto the greatest of them saith the Lord: for I will forgive their iniquity, and I will remember their sin no more.

The birth of Jesus arises on Plate 61 from an act of forgiveness, and the existence of Jesus is one act of forgiveness after another. The "perpetual re-enactment of the mystery of the Incarnation"[55] can occur with every birth, with every act of creation, with every act of art, because all those events are acts of forgiveness in which the divine takes shape in the fallen world. According to Altizer, the Incarnation, the Fall, and the Creation are particular moments of a single kenotic process, in which God empties Himself into the world.[56] As an extension of Altizer's insight, in the eternal world of Blakean Vision, Creation equals Fall equals Incarnation equals Crucifixion equals Resurrection equals Redemption, because a descent of the eternal into the fallen necessarily includes an ascent of the fallen into the eternal.

Ironically, and blessedly, the farthest that fallen vision can separate itself from divine Vision, the limit of contraction, is the apparent beginning of fallen history, Adam. And just as inevitably as Adam must appear, so must the separation of sexes occur, and so must Jesus eventually be born out of that separation. Not until man is completely blind to his own human divinity is any divine Incarnation necessary. In effect, man's deathly sleep far from the Divine Vision necessitates the appearance of Jesus. Contained in the very notion of a Fall is the movement of God into the world. Having separated from divine unity, man creates for himself a universe in which he cannot see God. He has in effect completely secularized existence. But when the universe is completely non-divine, then God Himself must be a man, and that is indeed what happens when Jesus is born. Fallen vision necessitates Creation which necessitates the Incarnation.

But the event of Incarnation, which is the ultimate extent of falling away, is also the beginning of re-unification. Just as the stars of the created universe are both evidence of the Fall and a merciful holding structure, just as the Bible as a work of art reveals the disastrous extent of the Fall and at the same time urges Regeneration, so the birth of Jesus occurs only because man has fallen so far, but at the same time it assures his re-unification with God. God becomes as we are so that we may become as He is. Once Jesus is born, then the Crucifixion is inevitable, for spirit which becomes completely flesh must die. And paradoxically, the Crucifixion, which is the most horrible moment in Christianity, is also the most celebrated and joyful, because it shows the final triumph of Eternity over fallen vision through apparent defeat. And all these events exist, not in historical time, once and for all, but in every moment of life.

Blake does not make the Crucifixion the central event of his Christianity, but the forgiveness of Jesus on the cross for those who have killed Him is central. The Incarnation of Jesus is an act of forgiveness in the deepest sense because it accepts completely the fallen world while at the same time transforming it through the resurrection which follows the crucifixion. The death of Jesus in the fallen world allows him to pass through the apparent limit of death into a resurrection that absolutely reverses the power of death and the Fall even while appearing to succumb to it. In Blake's theology there is no possibility of falling away irretrievably, for the merciful limits hold man in safety. The only thing that keeps him blind to the Divine Vision is Negation: doubt, despair, and abstract reason. Simply by expanding his perception he can achieve the unity which has been lost.

But the simplicity of the act does not lessen the agony of fallen vision or the difficulty of re-attaining total Vision. The change in perception is not simply a different way of seeing what is already seen, but is a totally different way of existing. It is not simply epistemology, but ontology. That is why any generalization is inadequate to explain the conversion which Blake urges. It does not come about except through total engagement with the fallen world to achieve the Eternity which exists within it. Once that Vision is achieved, all of existence is transfigured, and what were formerly perceived as separate parts of a difficult process can now be seen as joyous mental and spiritual warfare.

When Jerusalem misunderstands the process, she laments the hopelessness of her condition:

> My tents are fall'n! my pillars are in ruins! my children
> dashed
> Upon Egypts iron floors, & the marble pavements of Assyria;
> I melt my soul in reasonings among the towers of Heshbon;
> Mount Zion is become a cruel rock & no more dew
> Nor rain: no more the spring of the rock appears: but cold
> Hard & obdurate are the furrows of the mountain of wine &
> oil:
> The mountain of blessing is itself a curse & an astonishment:
> The hills of Judea are fallen with me into the deepest hell
> Away from the Nations of the Earth, & from the Cities of the
> Nations;
> I walk to Ephraim. I seek for Shiloh: I walk like a lost
> sheep
> Among precipices of despair: in Goshen I seek for light
> In vain: and in Gilead for a physician and a comforter.
> (79:1–12, E234)

But Jerusalem's despair contains within it a hopeful irony: if all merciful places have become identical with the enemy, then the enemy has become identical with mercy. And indeed that is the movement of Chapter 4 of *Jerusalem*. Fulfilling the movement of the eternal Jesus into the fallen world by reversing it, Chapter 4 traces the return of the fallen into the eternal. The agony and the despair are no less present and no less real, but they are inextricably bound with the powers of redemption. Jesus's descent into the fallen world, hinted at in Chapters 1 and 2 and made explicit in Chapter 3, brings the power of forgiveness which will transfigure the fallen into the eternal. In Chapter 4 the primary vehicle of that transfiguration is the reversal of Albion's turning away from Jesus on Plate 4 in Chapter 1. The agonies of the Fall will become peripheral visions, and all human forms will be identified through their Emanation Jerusalem.

The illustration on Plate 81 presents this basic truth in another way. Concealed behind her loins in mirror writing is a motto to which Gwendolen points:

> In Heaven the only Art of Living
> Is Forgetting & Forgiving
> Especially to the Female
> But if you on Earth forgive
> You shall not find where to Live.

This motto must be physically reversed to be read, and it must be conceptually reversed to be understood. As Erdman explains:

> The point of course is that the falsehood, as we come to expect of worldly wisdom, is only the truth turned inside out. It consists of the second half of Gwendolen's quatrain in mirror writing, 'But if you on Earth Forgive, You shall not find where to Live.' The true message? If you want to live in heaven, then start 'Forgetting and Forgiving,' which 'In Heaven is the only Art of Living'—the first half of the quatrain. (The Second half is almost completely painted out in E: Blake hiding the falsehood so that we are safe with the remainder read ironically or straight.) Her line of comment, 'Especially to the Female,' marks her false feminism.[57]

When Gwendolen sees that the infant she has nursed is a winding worm, she has no choice but to join Los in trying to shape it into something human. As does Vala, she shares Los's desires to form humanity, but she thought that the process needed war, sacrifice, and secrecy. Now, with Los, she begins "to form the Worm into a form of love by tears & pain" (E944).

Los is comforted in his pain, and states the knowledge which has often kept him going during his troubles:

> I know that I am Urthona keeper of the Gates of Heaven,
> And that I can at will expatiate in the Gardens of bliss.
> (82:81–82)

However, even with this knowledge, Los fears that if he immerses himself too much in the fallen world, he will lose eternity:

> But pangs of love draw me down to my loins which are
> Become a fountain of veiny pipes: O Albion! my brother!
> Corruptibility appears upon thy limbs, and never more
> Can I arise and leave thy side, but labour here incessant
> Till thy awaking! yet alas I shall forget Eternity!
> Against the Patriarchal pomp and cruely, labouring incessant
> I shall become an infant horror.
> (82:83–84, 83:1–5, E241)

Not realizing how close he is to seeing Albion awaken, Los here doubts the very principle of action which will help bring about the desired result: immersion in the fallen world. He knows that he cannot leave Albion in the tomb, subject to the concealments of Vala, but he forgets that complete descent into the horror of infancy is precisely what Jesus accomplished in Chapter 3. He fears that the constant work which is absolutely necessary may lead to the opposite of what he intends.

By His very birth Jesus becomes Satan; He does not simply enter the state, as some individuals do, but He actually becomes the state of Satan so that He can reverse Himself and put it off, thus freeing individuals. In a move similar to the one I described at the end of Pope's *Essay on Man*, individuals can participate in this power of the universal Jesus, not by appropriating to themselves the characteristics of Jesus, but by continually giving up any notions of universality. That is, any individual who thinks that he is a universal is denying his own individuality, the only place where universality resides:

> Los cries: No Individual ought to appropriate to Himself
> Or to his Emanation, any of the Universal Characteristics

Of David or of Eve, of the Woman, or of the Lord.
Of Reuben or of Benjamin, of Joseph or Judah or Levi
Those who dare appropriate to themselves Universal Attributes
Are the Blasphemous Selfhoods & must be broken asunder.
(90: 28–33)

Just as Eternity can be found only in the minute particulars of existence, so universality can be found only in particular individuals who continually break into pieces the blasphemous Selfhoods who try to appropriate universality to themselves.

The most pernicious appropriation of universality which Blake wishes to combat here is the narrowly orthodox notion of Jesus, which he has already severely undercut in Chapter 3, especially Plate 61. The orthodox Jesus removes Himself from humanity by being born of a virgin, thus quitting the field and allowing the cycle of Generation to crucify, enshrine, and neutralize Him, as in "The Mental Traveller." Such a Jesus must take on the Satanic nature of Generation and transform it into Regeneration. The question is, if all notions of universality, generalization, abstraction, and holiness are to be put off, then what is to take their place? Without a new mode of being, continual putting off would become an infinite regress.

Los's answer imitates the embrace of the details of the world which was the birth of Jesus:

...the Worship of God, is honoring his gifts
In other men: & loving the greatest men best, each according
To his Genius: which is the Holy Ghost in Man; there is no other
God, than that God who is the intellectual fountain of Humanity;
He who envies or calumniates: which is murder & cruelty,
Murders the Holy-one. (91:7–12)

And on an even simpler level, the individual can actually see Blake's Jesus, who will soon appear to Albion, immanently alive in others:

He who would see the Divinity must see him in his Children
One first, in friendship & love; then a Divine Family, & in the midst
Jesus will appear; so he who wishes to see a Vision; a

perfect Whole
Must see it in its Minute Particulars; Organized & not as thou
O Fiend of Righteousness pretendest; thine is a Disorganized
And snowy cloud: brooder of tempest & destructive War.
You smile with pomp & rigor: you talk of benevolence & virtue!
I act with benevolence & Virtue & get murdered time after time:
You accumulate Particulars, & murder by analyzing, that you
May take the aggregate; & you call the aggregate Moral Law:
And you call that Swelld & bloated Form; a Minute Particular.
But General Forms have their vitality in Particulars: & every
Particular is a Man; a Divine Member of the Divine Jesus.
(91:18–30)

In order to find the Jesus who will redeem us, we must see other individuals as individuals, minute particulars, and not as embodiments of any abstractions. Minute particulars are not atomistic parts which aggregate together into a whole, but each one is "a Divine Member" in that it fully participates in the divinity of Jesus. To see that Jesus, Los counsels us to look at others. We begin with our children (or, for the childless Blake, his artistic creations), and then extend that perception to all of the visible universe. This process is not based on a sentimental analogy, but on a hard-edged paradox: the apparent surface of our world is actually the center of the eternal world, but perceiving that reversal requires continual re-creation of the Self, in fact a continual annihilation of the Self, which is a false center, so that eternal existence can be constantly entered.

Notes

1. David Erdman, ed., *A Concordance to the Writings of William Blake*, 2 vols. (Ithaca: Cornell UP, 1967), 2:2181; 1:xii.

2. Qtd. in Deborah Dorfman, "Knowledge and Estimation of Blake during His Lifetime," in Adams, *Critical Essays*, p. 15.

3. 6 Feb 1818, Griggs, *Letters* 4:834, qtd. in Dorfman, p. 20.

4. Holmes, *Coleridge*, p. 49.

5. Coleman and Otto, *Introduction*, p. xi.

6. Peter Otto, *Constructive Vision and Visionary Deconstruction* (Oxford, 1991), p. 7.

7. Stephen Behrendt, *Reading William Blake* (NY: St. Martin's, 1992), p. 18.

8. Frye, *Fearful*, pp. 356–57.

9. I discuss the complexities of this couch/tomb in more detail in "Striving with Blake's Systems," in *Blake and His Bibles*, ed. David Erman.

10. *Blake's Night* (Belknap Press of Harvard UP, 1973), pp. 137–38.

11. Oxford: Oxford University Press, 1969.

12. Wagenknecht, pp. 4, 6.

13. Wagenknecht, p. 290.

14. Baton Rouge: Louisiana State UP, 1976, p. 2.

15. Schneidau, p. 302.

16. 2 vols., trans. Kendrick Grobel (NY: Scribners, 1951, 1955), 1:22.

17. *The Divine Imagination* (Metuchen, NJ: Scarecrow Press, 1972), p. 145.

18. Altizer, p. 194.

19. Bultman, *Theology*, 1:19–21. As this paradoxical eschatology permeates *Jerusalem*, it is worthwhile to note some other formulations by Bultmann:

> According to the New Testament, Jesus Christ is the eschatological event, the action of God by which God has set an end to the old world.... It is the paradox of the Christian message that the eschatological event, according to Paul and John, is not to be understood as a dramatic cosmic catastrophe but as happening within history, beginning with the appearance of Jesus Christ and in continuity with this occurring again and again in history, but not as the kind of historical development which can be confirmed by any historian.... although the advent of Christ is an historical event which happened 'once' in the past, it is, at the same

> time an eternal event which occurs again and again in the soul of any Christian in whose soul Christ is born, suffers, dies and is raised up to eternal life.... every instant has the possibility of being an eschatological instant and in Christian faith this possibility is realised.... In every moment slumbers the possibility of being the eschatological moment. You must awaken it. (Bultmann, *History,* pp. 151–55)

From the beginning to the end of *Jerusalem* the call of the poet is to the reader and to all of humanity to awaken to the eschatological possibility which slumbers within each moment. And the triumphant conclusion is the realization of that moment, the absolute end to history which paradoxically can not be confirmed by the historian, the absolute entry into Eternity, which can not be perceived by fallen vision.

20. Berkeley: Univ. of California Press, 1972, pp. 10–12.

21. Peter Fisher, *The Valley of Vision* (Univ. of Toronto Press, 1961), p. 47.

22. For Blake's accusation that Plato distorted Socrates's teachings, see "A Vision of the Last Judgment," E554.

23. Damrosch, p. 114.

24. Damrosch, pp. 175, 238.

25. Damrosch, pp. 240, 380, 370.

26. Damrosch, p. 371.

27. Fisher, p. 188.

28. I discuss the question of "system" more fully in my "Striving with Blake's Systems," in *Blake and His Bibles.*

29. *Marriage of Heaven and Hell,* plates 4, 12; *Song of Los,* plate 3; *Laocoon; Europe,* plate 12; and two examples in the annotations to Watson.

30. Frye, *Fearful,* p. 416.

31. For an attempt at a more rational, less paradoxical explanation of Blake's inconsistent God, see H. Summerfield, "Blake and the Names Divine," *Blake: An Illustrated Quarterly,* 57 (Summer 1981): 14–22.

32. Blake is following the line of Milton, who refused bondage to rhyme in his preface to *Paradise Lost*:

> Rime [is] no necessary Adjunct or true Ornament of Poem or good Verse.... This neglect then of Rime ... is ... an example ... of ancient liberty recover'd to Heroic Poem from the troublesome and modern bondage of Riming. (*Complete Poems*, p. 210)

33. Mitchell, "Visible," p. 75.

34. This is the only plural use, "types"; the other four, singular, which refer simply to printing, are found in a picture title and in letters.

35. The best account of this labor is found in Robert Essick, *William Blake, Printmaker* (Princeton UP, 1980), from which the following simplified information is taken.

36. For a discussion of writing backwards so that the reversed design will read correctly, see Essick, *Printmaker*, pp. 89–92.

37. For one way to understand this reversal, see Frye, *Fearful*, p. 383.

38. For a fuller discussion of this dynamic, see the following chapter.

39. *Blake and the Assimilation of Chaos* (Princeton UP, 1978), pp. 10–15.

40. W.J.T. Mitchell rescues Urizen even more firmly from the derision of traditional Blake criticism:

> Urizen is no doubt sometimes employed as a figure of English reaction in the late 1790s, but it is also clear that in *The Book of Urizen* (1794) Blake represents him as a revolutionary, utopian reformer who brings new laws, new philosophies, and a new religion of reason. (Mitchell, "Visible," p. 58.)

41. Gallant, pp. 76–77.

42. *Acts of Inclusion*, p. 139.

43. Karl Kroeber, "Delivering *Jerusalem*," *Blake's Sublime Allegory*, ed. Curran and Wittreich (Madison: Univ. of Wisconsin, 1973), p. 366.

44. Frye, *Fearful*, p. 357.

45. Except in a letter to Josiah Wedgwood, and even there it refers to a series of identical engravings, E770.

46. Cooke, *Acts of Inclusion*, p. 219.

47. Erdman, *Blake: Prophet Against Empire*, 3rd ed. (Princeton UP, 1977), p. 469.

48. *Poems of William Blake* (1905; rpt. London: Routledge, 1969), p. xviii.

49. Wagenknecht, p. 273.

50. Prickett, *Words*, pp. 163–64.

51. Damrosch, p. 168.

52.

> Blake ... intensified the process of defamiliarization by an ironic undermining of the accepted.... For Blake, irony was a necessary precondition of vision as etching acid is necessary to produce ultimately the "illumination" of the print. (John Howard, *Infernal Poetics* [Rutherford: Fairleigh Dickinson Univ. Press, 1984], pp. 24–25)

53. *The Divine Image* (Rome: Edizioni di Storia e Letteratura, 1950), p. 81.

54. Altizer, p. 201.

55. Bottrall, p. 14.

56. Altizer, p. 105.

57. *The Illuminated Blake*, p. 360.

Chapter 8
Monos o Iesous: The Transfiguration of the Bible in *Jerusalem*

> The full meaning of the Incarnation is that the Incarnation is a dual and dialectical process whereby God empties Himself of Himself and becomes man and man empties himself of his historical particularity and his individual selfhood and becomes God: 'Therefore God becomes as we are, that we may be as he is.'... Spirit is this eternal movement of absolute self-negation.[1]

Nestled in the waxing crescent of the moon at the top of plate 4 of *Jerusalem* appears the transliterated Greek inscription, "Monos o Iesous." Etched in white line against a black background, the phrase is linked by a star to the letter "J" in the title "JERUSALEM." A floating female figure, identified by David Erdman as Jerusalem herself,[2] points to the words. Commonly translated as "Jesus only," the phrase holds a position that invites us to see it as a motto for the entire poem. Thomas Altizer assigns it even more importance: "the motto which Blake in fact gave *Jerusalem*, 'Jesus only,' is the key to the ultimate meaning of his vision as a whole."[3]

"Monos o Iesous," in its simultaneous allusions to two Bible scenes, the woman taken in adultery and the transfiguration, urges us toward two simultaneous perspectives on Jesus, his transcendence and his immanence, and, like *Jerusalem* itself and Blake's entire œuvre, urges us to hold these two apparently mutually exclusive pictures in our mind simultaneously. Transcendence occurs primarily in the transfiguration, in which Jesus hovers above man. Immanence occurs primarily in the forgiveness of the woman taken in adultery, by which Jesus fully enters into the world and accepts it in full participation in everyday life. Transfiguration and forgiveness fit together in a coincidence of opposites to exemplify and enact the contrary vision of Blake. Furthermore the two are inseparable because the transcendence of transfiguration is achieved through the immanence of forgiveness. Indeed, even though the two biblical incidents emphasize contrary aspects of divinity, each contains within it both transcendence and immanence. Each combines acceptance and rejection of the fallen world.

Thus this motto exemplifies Blake's solution to the problem of oppositions. Standing first in his most comprehensive completed project, his greatest poem, *Jerusalem*, "Monos o Iesous" embodies and introduces the simultaneous transcendence and immanence that is the essence of Blake's vision. In that coming together of transcendence and immanence all other oppositions interpenetrate. *Jerusalem* becomes the fulfillment of the theory of *coincidentia oppositorum* put forth by Nicolas Cusanus. Although there is no evidence that Blake ever read Cusanus, the similarities in their thought and the ways that Blake's *Jerusalem* fulfills the goals of *Of Learned Ignorance* make the juxtaposition worthwhile.[4] Just as Blake fulfills the Bible by taking it beyond its own limits, so he does with the *coincidentia oppositorum* of Cusanus. For Cusanus left the coincidence of opposites in a mystery: as learned as we might become, we must always remain ignorant. Alexander Pope maintained that mystery: in *An Essay on Man* we must submit and accept that, "Whatever is, is right." Samuel Taylor Coleridge tried to envisage the simultaneous existence of divine and human, but, as we have seen in his lines to Berengarius, he never could. Mary Shelley urged a double vision, with self-abnegation working against rampant egotism, but *Frankenstein* reveals mostly the gloomy failure of that struggle. Percy Bysshe Shelley pierced further into the mystery than any of those, but the triumph of *Prometheus Unbound*, which itself is stopped short by the mystery of imagelessness, receded before the problems of *The Triumph of Life*.

In all his work, William Blake struggled with the vicious circles epitomized in "The Mental Traveller." In *Milton* and *The Four Zoas* he made heroic efforts to come to terms with the coincidence of opposites. Only in *Jerusalem* did he fully succeed. "Monos o Iesous" is the emblem to that success.

The words "Monos" and "Iesous" occur together in only two passages in the Greek New Testament. In John 8:9, during the story of the woman taken in adultery, they occur almost exactly as in the heading in *Jerusalem*, allowing for orthographical errors: "Monos o Iesous." In the account of the transfiguration in Luke 9:36 the words "Iesous Monos" occur. (The transfiguration accounts in Matthew and Mark do not use the word "monos.") Michael Tolley[5] credits W.H. Stevenson[6] with noting the two possible Bible references, but labels both as doubtful. Alicia Ostriker writes that both references have relevance to *Jerusalem*.[7] Joseph Wicksteed suggests that the words might refer to the transfiguration,[8] but in response Anne Mellor maintains that the occurrence of the exact words only in John makes that reference the more likely one.[9] Johnson and Grant[10] simply translate the Greek phrase, "'Jesus alone' (John 8:9)," thus implicitly preferring only the allusion to the woman taken in adultery. The two major texts of Blake, Erdman's and Keynes's, are silent on the subject.

There is no doubt that Blake was familiar with the Greek New Testament. Although he may be exaggerating, we need not doubt his essential claim in a letter to James Blake: "I read Greek as fluently as an Oxford scholar & the Testament is my chief master."[11] There is also no doubt that he was more than familiar with both the story of the woman taken in adultery and the story of the transfiguration. We have not only the general evidence of his thorough knowledge of the Bible but also the specific evidence of watercolor illustrations of both episodes.

The allusion to John 8:9 is unmistakable; the words are exact and the meaning is clear: the forgiveness that Jesus shows toward the woman taken in adultery, the same forgiveness that blossoms and flows between Joseph and Mary on Plate 61 of *Jerusalem*, is one of the most obvious themes of the epic. For example, in the introduction on the plate preceding "Monos o Iesous" the poet declares, "the Spirit of Jesus is continual forgiveness of Sin."

Less obvious, but equally unmistakable, is the allusion to Luke 9:36. As I shall show, both the words and the illustration on Plate 4 clearly echo Luke's transfiguration scene. Furthermore, on Plate 96 when Albion finally awakes from his sleep and sees the Jesus whom he has turned away from on Plate 4, a scene of transfiguration, analogous to that in the Bible, is completed. Thus a transfiguration scene frames the whole of *Jerusalem* and informs its every plate.

> At one hundred plates *Jerusalem* is Blake's longest illuminated work, and yet underlying its great philosophical and mythological complexity is a startlingly simple premise: the entire poem presents from a variety of perspectives the single act of Albion's awakening to consciousness and his concomitant full individuation.[12]

By reference to the two Bible stories, and to the transfiguration echoes in *Jerusalem*, and to the two watercolor illustrations by Blake on the subjects of the transfiguration and the woman taken in adultery, I shall show how Blake combines both stories together into a new relationship between the human and the divine which is explored and urged throughout the poem.

The story of the transfiguration as told in the gospel of Luke clearly emphasizes the replacement of the old law by the new and the majestic transcendence of the son of God:

> he took Peter and John and James, and went up into a mountain to pray. And as he prayed, the fashion of his countenance was altered, and his raiment was white and glistering. And, behold, there talked with him two men, which were Moses and Elias: Who appeared in glory, and spake of his decease which should accomplish at Jerusalem. But Peter and they that were with him were heavy with sleep: and when they were awake, they saw his glory, and the two men that stood with him. And it came to pass, as they departed from him, Peter said unto Jesus, Master, it is good for us to be here: and let us make three tabernacles; one for thee, and one for Moses, and one for Elias: not knowing what he said. While he thus spake, there came a cloud, and overshadowed them: and they feared as they entered into

> the cloud. And there came a voice out of the cloud, saying, This is my beloved Son: hear him. And when the voice was past, Jesus was found alone [*monos*]. And they kept it close, and told no man in those days any of those things which they had seen. (Luke 9:28–36) [13]

Terence Hoagwood notes that the both the text and the illustration on plate 4 are replete with contraries:

> The first page of the first chapter of *Jerusalem* confronts the viewer with a vision of contraries before he begins to read the text. Right and left male, upward and downward sequences, and materialism (with its moral law) and spiritual rebirth are all set in clear contrast, before the poem even begins.[14]

Both the text and the illustration on Plate 4 are strikingly analogous to the transfiguration scene in Luke. First, in the text, the poet begins by declaring that his subject is a "Sleep" and an "awaking" from that sleep. In Luke the apostles fall asleep while Jesus prays and do not wake until he is transfigured and engaged in conversation with the two prophets. Even when the apostles awaken from their sleep (a sleep, like Albion's throughout *Jerusalem,* which is evidence of their lack of understanding, of their turning away from vision), they still do not understand the situation. Peter wants to build three tabernacles instead of worshiping only Jesus. The apostles need the direct voice of God to change their way of thinking; they need to replace the old law of tabernacles with the new law of Jesus only. They are awakened from sleep, and a voice from above corrects and inspires them. Similarly, in *Jerusalem*, the Savior's voice, which is "over" the poet, calls for an awakening: "Awake! awake O sleeper of the land of shadows, wake! expand!" The voice then goes on to correct mistaken impressions: "I am not a God afar off, I am a brother and friend." The direct voice of Jesus insists on replacing old concepts with new. Just as the voice of God in Luke corrects the misunderstanding of the recently awoken apostles and inspires them to regard the proper vision of Jesus, so the voice of Jesus in *Jerusalem* corrects the sleeping Albion and inspires the poet toward imaginative Vision. And in both Luke and *Jerusalem* the voice which speaks to the few is understood to be speaking also to the many, to all readers of the text.

Once the closeness of the analogy has been noted, the differences become crucial. For even while Blake is imitating the story of the transfiguration, he rewrites it in his own terms. Blake the poet *sees* the source of his inspiring voice—"I see the Saviour over me"—while the apostles in Luke see only the cloud from which the voice of God broadcasts. The apostles do not see clearly and do not understand clearly, not only because they are sleepy but also because God is coy. In fact the God of the scene in Luke is little improvement over the Old Testament God who sheltered his face from view; the God who declares the supremacy of his son Jesus is still the unapproachable Yahweh of Moses. Not only can Blake see his dictating Saviour, but the voice itself promises a deeper unity with the poet, total interpenetration of the human and the divine: "Within your bosoms I reside, and you reside in me."

In the New Testament transfiguration story God is the supreme invisibility, distant and beclouded; the apostles are inferior sleepyheads who worship in the wrong way, far beneath their God. They must look up to see him and at best they see a cloud; even after they are corrected, their state of blind sleep has been raised only to a state of foggy obeisance. Blake re-interprets the story: in *Jerusalem* the Savior is visible and accessible; the poet, as well as all humanity, needs to wake up and see clearly what is possible beyond ordinary limits. As Jesus declares later in *Jerusalem*:

> ... I am the Resurrection & the Life.
> I Die & pass the limits of possibility, as it appears
> To individual perception. (62:18–20)

The Vision is thus made available at the beginning of the poem, but it is not until nearly the end that Albion does awaken. When he first has the opportunity on Plate 4, he turns away and falls into sleep/death, as he declares his opposition to Jesus's vision:

> Phantom of the over heated brain: shadow of immortailty!
> Seeking to keep my soul a victim to thy Love!
> ...
> By demonstration man alone can live, and not by faith.
> ... here will I build my Laws of Moral Virtue.
> (4:24–29)

Hoagwood precisely notices the reversal that Blake has set up:

> [Albion] has assembled the same principles and contexts that Christ had assembled above, but in reverse perspective; in this way, the text of plate 4 reproduces the patterns of contraries and reversal that characterize the preceding designs.[15]

In effect the whole of *Jerusalem* performs a transfiguration and finally breaks through the cloud of obfuscation in which Albion is trapped, blind to Vision as the apostles are blind to the true nature of Jesus. The poem is long and repetitive because the process meets recalcitrance, especially in the person of the Spectre.

But the deepest intricacies and most serious wit of Blake's conception here, important for the rest of *Jerusalem* and for all of Blake's notions of religion and art, lie in his new treatment of the middle term, Jesus. In Luke the voice of God in a cloud above speaks to the human apostles far below on the subject of the middle term, Jesus the mediator. However, Luke's Jesus does nothing active; in his glory he is a passive emblem of the power of God, merely an extension of the father. He remains distant, strikingly different from the humanity in which he supposedly partakes as a combination of man and God. And he is further separated from humanity in that the apostles do not tell anyone else about the miracle ("they kept it close"). This Jesus is to be worshipped in fear and mystery and secrecy, as if he were identical with the mysterious Jehovah of the Old Testament.

In the text of *Jerusalem* 4 the voice of Jesus above speaks to Albion and to the poet on the subject of divine-human love. In particular Albion is urged to repair the divisions with brothers, fathers, sons, nurses, sisters, and daughters—in other words to become fully human again. This advice is virtually identical to a method which Los will later declare on Plate 91:

> He who would see the Divinity must see him in his Children
> One first, in friendship & love; then a Divine Family, & in
> the midst
> Jesus will appear. (91:18–20)

But Albion must go through the horrors of eighty-seven more plates before he acts on that advice. Albion is further urged to repair divisions with his Emanation and with the Savior himself.

Here the middle term which the voice from above explains is not an emblematic glistering mystery, but the very power of the divine human unity itself.

In Luke the apostles are told to conceive of the transcendent power in a new way; in *Jerusalem* Albion is told to see his own divine humanity in a new way. In Luke the figure of Jesus stands as an emblem of the relationship between man and God: God shines and man fears. Blake places before his reader an emblem of the divine creativity of human love in the form of his poem *Jerusalem*. Completely visible in its minute particulars, *Jerusalem* becomes the mediator, taking the place of Jesus in the analogy between the transfiguration and *Jerusalem*. If we can read *Jerusalem* (or, by implication, any work of art which appeals to the spiritual understanding) with expanded Vision, then we can participate in a transfiguration on a more intimate level than did Peter, John, and James. We can see the words and illustrations which are simultaneously fallen and Visionary; and we can see the Jesus who is both God and man. And that reading is itself a transfiguration.

It is ultimately a transfiguration that Blake is creating and displaying in *Jerusalem*, transfiguring the Bible, himself, and all of existence, through the power of creativity and urging us as readers to do likewise. The three terms of humanity, divinity, and mediator (who combines both) come together in a coincidence of transcendence and immanence. For the Jesus who hovers over the poet and the Jesus who is one with the poet are the same; in fact in *Jerusalem* the God who gives the word from Sinai and the Jesus who forgives sins are the same.

The message of the poet who claims to follow Jesus while he simultaneously usurps his position is that each individual must awaken from his mistaken worshiping. The only way to stop worshiping falsely is to recognize the creativity which is the presence of God in each man. Thus Blake does not urge us to give up a transcendent deity, but he does insist that the divine is as much immanent as it is transcendent. The useful distinction to make is not between transcendence and immanence, just as in Plate 10 the useful distinction is not between good and evil. Those contraries should both be fully embraced. The useful distinction is the one between creative individuality, which leads to total human and divine Vision, and secretive vengeance, which leads to chastity, warfare, and punishment.

Blake transfigures Jesus into a creative power more intimate than priestly separation and secrecy allow for. And to be more intimate Jesus must expand his role, just as he commands Albion and the poet to do ("awake, expand"). In fact Blake's Jesus, in the form of Blake's creativity, expands to fill all three roles in the transfiguration: God, intermediary, and human; thus the basic structure of the transfiguration scene is maintained even while it is subverted and denied. "Jesus only" takes on a transfigured meaning. Thus the transfiguration scene, the Bible, Blake, and perhaps the reader are transfigured into new Vision. To be the best possible readers of *Jerusalem* we must transfigure it by our own powers of creativity, just as it does to the Bible.

Like the text on Plate 4, the illustration at the top of Plate 4 is also strikingly similar to the transfiguration scene in Luke (see figure 1 on p. 232). At the top of the plate are the words which must be heard ("Monos o Iesous"); in the middle is a cloud which simultaneously obscures and reveals; at the bottom are three figures whose vision is imperfect. On top of a mountain which contains the engraved text sit three figures, analogous to the apostles in the Bible account, who cannot see the words above them. The figure on the left gazes upward in the right direction, but his vision is obstructed by the cloud which envelops the title "JERUSALEM." Even if he could see through the cloud, the crescent moon would cut off the word "Iesous" from his vision and leave "Monos" only. The figure on the right also gazes upward, but his line of sight is far off the mark. The middle figure appears to be restraining both men's attempts to see, and she stares straight at the reader as if to hypnotize him and becloud his vision. However, the line of her outstretched right arm (pointing to the reader's left) leads directly to a chain of three flying children who circumnavigate the cloud befogging "JERUSALEM." There might thus be a way indicated: go around. The children are lifted upward by the floating Jerusalem herself, who points directly to the essential words at the top of the plate, "Monos o Iesous."

Erdman says that the scene below the title on Plate 4 is a misjudgment scene.[16] Both Bible references give us scenes of people misjudging and being set straight by a divine voice. Just as Peter, James, and John mistakenly want to erect temples to Moses, Elias,

Figure 1. *Jerusalem*, plate 4
(Department of Printing and Graphic Arts,
The Houghton Library, Harvard University)

and Jesus, so the Albion figures under the sibyl's hands want to remain under her grasp. But the female figure above, like the female babe in "The Mental Traveller," has completely escaped the limits of the scene below. Just as the words of God in the transfiguration scene proclaim Jesus to be the beloved son and the Bible text proclaims him to be alone, so the words at the top of the page inform and correct the mistaken judgment below them. These words, which echo the descending words of Jesus etched in the plate below are soon denied by the Spectre who stands "over" Los (6:4 and illustration on Plate 16) and provides the negative of the words of God and Jesus. But the wings of the Spectre in the illustration and his words are designed to hide the true light from the poet. When he looks up on Plate 6 all he can see is the Spectre. When the left figure on Plate 4 looks up, he sees the title "JERUSALEM" floating in the clouds, a few stars, and the word "Monos." The word "Iesous," which Jerusalem can see clearly and to which she points, is hidden from the figure's view by the moon and the cloud.

Blake's watercolor illustration of the transfiguration (see figure 2 on p. 234), painted as part of the Butts series of Bible illustrations, is built on a structure similar to that of the illustration on Plate 4. Jesus stands glowing in the center, dividing most of the picture vertically in half. Below his feet crouch Peter, James, and John, who have accompanied Jesus up the mountain. The two in the bottom corners gaze up, as do two figures in *Jerusalem* 4, at the face of the transfigured Christ, while the third, directly under the feet of Jesus, hides his face in his sleeve. The two gazing up appear to have just awoken and the third seems still asleep. The round, almost petaled top of the sleeper's head becomes a pedestal for the feet of Jesus and provides a contrast with the glowing face of Jesus at the top of the illustration. The humbled posture of the three apostles makes them appear to be part of the mountain; in fact, since no part of the actual mountain is visible, they seem to be the mountain. Kneeling in symmetrical bows appear Moses on Jesus's right (our left) and Elijah on his left. Their hands in a position for prayer, they too gaze up piously at Jesus's face. Here Blake has altered the Bible account which reports that Moses and Elijah stood and spoke with Jesus but does not say that they worshipped him. Thus Blake makes his transfigured Jesus even more lordly, even more transcendent than does Luke.

Figure 2. The Transfiguration
(Victoria & Albert Museum)

The lower garments of Moses, Jesus, and Elijah curl sinuously away from their feet, making them appear almost to float in the air, and separating them from the lower apostles. But what is most interesting is Blake's addition of two bearded patriarchal faces, with large wings, hovering in the background, one on each side of Jesus's face, at a slightly lower level. The appearance of these two characters does not simply alter the biblical text, but also adds a completely new dimension to it. At first glance, in an orthodox reading of the painting—and the painting certainly lends itself to being read as straightforward praise of this exalted Jesus—these two figures are simply two angels of God presiding over the ceremony. No such angels are mentioned in any of the accounts of the transfiguration, but they could easily be a pictorial representation of the invisible God who speaks in the scene. Indeed the picture perhaps encourages us to read the three faces across the top as somehow forming a divine trinity, set off from humanity. The three humans in the design, relegated to the bottom level of the painting where they cringe in bafflement, form an earthy ground on which the exalted Savior can display himself. Thus we have three divine faces across the top and three human ones across the bottom, the middle human, however, being not a face at all but the top of a head which looks like a vegetated flower. Across the horizontal center of the picture the two patriarchs bow, their faces level with the waist of Jesus. At all three horizontal levels Blake may be mocking the orthodox reading of the transfiguration: on the grounds that it makes Jesus into a ghostly unapproachable deity with no humanity, on the grounds that it causes prophets to humble themselves, and on the grounds that it denigrates men into footstools. However, despite the possible mockery, the watercolor retains a dignity and a splendor which cannot be dismissed by simply seeing the picture as ironic. The mockery may be there, but the majesty certainly is. A similar doubleness is useful in interpreting Blake's most famous painting: the godly Urizen bending from the clouds to measure or create the world with his compasses. Blakeans in the know usually read that picture ironically, but surely the transcendent majesty of the god remains, no matter what mockery exists.

The transfiguration and the woman taken in adultery present opposing views of Jesus. In the transfiguration, Jesus is separated from man, left alone as the only Son of God. In the story of the woman taken in adultery, Jesus lowers himself both physically and spiritually, as he implies that none of us is free of sin:

> And the scribes and Pharisees brought unto him a woman taken in adultery; and when they had set her in the midst, They say unto him, Master, this woman was taken in adultery, in the very act. Now Moses in the law commanded us, that such should be stoned: but what sayest thou? This they said, tempting him, that they might have to accuse him. But Jesus stooped down, and with his finger wrote on the ground, as though he heard them not. So when they continued asking him, he lifted up himself, and said unto them, He that is without sin among you, let him first cast a stone at her. And again he stooped down, and wrote on the ground. And they which heard it, being convicted by their own conscience, went out one by one, beginning at the eldest, even unto the last: and Jesus was left alone [*monos*], and the woman standing in the midst. When Jesus had lifted up himself, and saw none but the woman, he said unto her, Woman, where are those thine accusers? hath no man condemned thee? She said, No man, Lord. And Jesus said unto her, Neither do I condemn thee: go, and sin no more. (John 8:3–11)

The Jesus who humbles himself into the pains of the world and the Jesus who speaks to Blake and to Albion and to us all is the Jesus of the forgiveness of the woman taken in adultery. Blake's watercolor of this scene (see figure 3 on p. 237) provides a sharp contrast to the image of Jesus presented in the illustration of the transfiguration. As with the transfiguration, it is useful to divide the illustration into three sections, this time from right to left instead of top to bottom. On the right stands the accused woman, dressed in a flowing white robe against a dark background. Barefoot, her hands bound behind her back, she inclines her head to read what Jesus is writing in the sand with his finger. Her hair falls, but although he is bending more than she is, his hair does not fall. Also barefoot and wearing a white robe similar to the woman's, Jesus bends his back into a rounded arch to reach the ground. The curve of his back is an ironic echo of the arched doorway which appears behind him at the top of the picture through which seven accusers shoulder each other in their

Figure 3. *Woman Taken in Adultery*
(Purchased, 1890; Courtesy, Museum of Fine Arts, Boston)

haste to escape. As with the middle figure at the bottom of the watercolor of the transfiguration, we do not see their faces. To emphasize immanence instead of transcendence, whereas in the illustration to the transfiguration the face of Jesus is the highest thing in the picture, here it is the lowest of the nine heads in the picture. It is in fact level with the genitals of the woman, and pointing straight toward them as if to emphasize the forgiveness of her particular sin.

Along the bottom of the illustration are featured from left to right the bare right foot of Jesus, his hand writing on the ground, and the bare left foot of the woman. Alternating with these three bare extremities are two dark sandaled fleeing feet of two of the accusers. Thus, despite the obvious separations between woman and Jesus, between accusers and two main figures, the feet and hands on the ground form an alternating pattern of unity, just as does the lesson of Jesus's injunction to cast the first stone.

Of course the Jesus who writes in the sand (we never know what he writes) is an echo of the Blake who engraves with his pen the words and pictures of *Jerusalem*.[17] The pointing finger implicitly contrasts with the pointing fingers of the accusers on Plate 93 and echoes the finger of the female figure on Plate 4. In all cases the contrast is made clear. The pointing fingers of Jesus, Jerusalem, and the engraving tool of Blake all point in alternative directions to the old law, punishment, and sleepy failure to see.

The old law called for the execution of an adulteress (see Lev. 20:10), but the new law of Jesus does not allow for such easy separation of the sinner and the righteous. The scribes and the Pharisees see the situation in very clear terms: the woman is an adulteress and must be punished according to the law.

> As a totality, the self is by definition always a *complexio oppositorum*, and the more consciousness insists on its own luminous nature and lays claim to moral authority, the more the self will appear as something dark and menacing.[18]

They have been trying, in the chapter of John preceding this incident, to discredit Jesus and the belief that he was the Messiah. They fully expect him to pardon the woman and thereby expose himself to arrest for breaking the law of Moses. If instead he does condemn her, then they have trapped him into being just as vindictive as they are. But Jesus knows how to break the law without really breaking it.[19] At first he writes on the ground, as if he does not hear the accusers. Biblical scholars are not much help in trying to guess what Jesus might have written, but in any case it is clear that his writing is not even the kind of response the scribes and Pharisees expect.

Then when Jesus speaks, his answer is even more unexpected: he confronts the accusers with their own nature as sinners. The self-righteous have projected all their feelings of guilt and sin on to another person, a kind of scapegoat (an honored tradition in Mosaic law). Similarly, when we as readers turn to Blake, he does not speak to us in the ways we might expect; his forms are asyntactical and idiosyncratic, his illustrations puzzling and impossible. His morality is not moralistic. He confronts us with our own accusations, or more exactly he confronts our selfhoods and forces them to turn tail and run. "Each Man is in his Spectres power / Untill the arrival of that Hour / When his humanity awake / And cast his Spectre into the Lake."[20] The accusers in Blake's illustration scatter just like the self-righteous selfhood scatters when confronted with the identity between itself and a sinner. Throughout *Jerusalem* Blake calls Jesus "the friend of sinners" and forces the reader to confront his own sinfulness.

Both the transfiguration and the story of the woman taken in adultery insist that we see existence in a new way: transcendent in the first story, immanent in the second. "Monos o Iesous," in making these two opposites coincide, insists that we confront Blake's Christian art.

The essence of Christianity, and therefore of the Bible from a Christian point of view, is also its most fundamental paradox: the coincidence of divine and human in Jesus Christ. As we have seen in Cusanus and Coleridge, that coincidence is also central to their conceptions of existence. All the rest hangs from that basic belief. The essence of the Bible, from the Creation through Yahweh's covenant with Israel, from the birth of Jesus through the letters of Paul, from Eden through the new Jerusalem, is the intersection of the divine and the human, the temporal and the eternal. The essence of Blake is the imaginative Vision which sees the eternal and the temporal whole, without sacrificing either one: Eternity in a moment, Infinity in a flower.

In a Christian reading of the Bible, the coming of Jesus in the New Testament is the fulfillment of the relationship between man and God in the Old Testament. At the same time the new law destroys and removes the old. From that perspective an analogy with the two parts of the Bible itself can be helpful in understanding the use of the Bible in *Jerusalem*: *Jerusalem* stands in relation to the Bible as the New Testament stands to the Old. To most Christians, following the teachings of Paul, the New Testament removes the burden of the old law; it destroys the law, replaces it. At the same time, most of those same Christians, following the teachings of the writer of the epistle to the Hebrews, see the New Testament as the fulfillment of the Old. Just as Jesus comes both to fulfill the law as the promised Messiah and to overturn it as the Son of God with a whole new plan of salvation, so Blake comes both to fulfill the Bible and to overturn it. As Margaret Bottrall succinctly puts the case: "Blake called himself a christian; his commentators point out that he was a heretic; both are speaking the truth."[21] Altizer gives this paradox its ultimate formulation; he calls Blake "the first Christian atheist."[22]

The New Testament breaks the limits of the Old Testament by extending them to their logical conclusions, by changing their terms, and by analogizing them into something higher, all the while respecting them as the foundation and the necessary prologue to Christianity. Blake uses several variations of these techniques: he extends the limits of the Bible when he rewrites the story of Joseph and Mary; he redefines Jerusalem, Canaan, Reuben, and a whole host of biblical places and persons; and he transfigures the nature of Jesus; all the while honoring the Bible as the ultimate model of the kind of art he is undertaking. Just as the New Testament includes the Old, so Blake's Jerusalem includes the Bible.[23]

From the Christian point of view, the Old Testament anticipates the New in that its history leads up to the New, in that its typology is a shadow of the New, and in that its morality is a dimmer version of that of the New.[24] From Blake's point of view, the Bible anticipates the artist/Christian in that it contains the entire context of time and space in which the artist lives and works, in that its language is often allegorical of spiritual situations, and in that its morality must be transcended.

In short, Blake imitates the Bible by treating it in the same way that the New Testament treats the Old: fulfilling and reversing. The paradox is unmistakable: to fulfill is to reverse; to reverse is to fulfill.

Thus in the New Testament stories of the transfiguration and the woman taken in adultery, and in Blake's watercolor paintings of the two scenes, we receive two very different pictures of Jesus. In the transfiguration he is transcendent, unapproachable, and (as exaggerated in Blake's illustration), susceptible only to worship and the humiliated adoration of humanity. In the woman taken in adultery he is forgiving, proud to the self-righteous, humble to the sinner, and the writer who sets men and women free from the tyranny of their spectre-accusers. It is appropriate that Blake should introduce the title page of *Jerusalem* with a motto that recalls both these incidents, and indeed turns them into one, or makes it impossible for us to separate them from each other. One Jesus comes to uphold transcendence, to inherit his place from the Old Testament God. The other comes to overturn the rigidities of the law, to become fully human. The Jesus of the New Testament does both; so does the Jesus of Blake; and so does Blake himself in his rewriting of the Bible. Like Jesus, he cannot be fully understood if either aspect is denied: he both carries on the tradition of the Bible and overturns it. As on Plate 99 the two principles embrace—the God who disperses law (in this case the new law of Jesus) and the human (the sexually sinning female). And so the transcendent unapproachable God and the indwelling human Jesus are one.

Much of Blake's teaching works the same way as the teaching of Jesus with the accusers of the woman taken in adultery. Certainly the Bible story can be read as a simple moral lesson: don't judge others. But the accusers (and of course the woman herself) experience something quite different from a platitude. They can not take refuge in the complacent generalizations that readers can sometimes sink into. Instead of being told the lesson that we can infer from the story, that no one is free of sin, they are confronted directly with that knowledge and with the eyes of Jesus and the woman. And the moment of knowledge and recognition conflicts with the action that the accusers are in the middle of taking. Stones mentally poised above their heads, they hear the words of Jesus (and presumably read the ones he writes on the ground, ironically lost to us in the written transmission of the gospel): accusation of other and accusation of self clash within them. Trying to trap Jesus in the paradoxes of the Law, they find themselves trapped in the paradoxes of a larger, more inclusive law, the law of the *coincidentia oppositorum*, the law of the illogical unity of mutually exclusive forces, the coming together of the centrifugal transfiguration, thrusting God out from man, and the centripetal forgiveness of the woman taken in adultery, joining God with man. Unable to respond physically, unable to respond logically, unable to respond spiritually, they run away. Similarly, Blake's text again and again forces us to see the opposing factions within ourselves, because they are displayed so vigorously in his poetry. Most readers therefore run away, either by refusing to read him at all or by pigeon-holing his message into a platitude.

The reader of Blake who does not run away can not contemplate the text in comfortable, smug, self-righteous tranquility, as he might Alexander Pope or even the Bible. Instead he has to confront the Eternity of Blake's minute particulars with every sweep of the eyes across the words and illustrations. Blake's simultaneous imitation and rejection of all experience, and of the Bible in particular, teach us a new way to read and a new way to live, a way that demands awareness of how our ordinary understanding distorts everything from religion to sex, from sight to philosophy.

Whichever side of a question a reader is on, for example in the matter of sexuality and the body (horrible trap or imaginative release) or unity and individuality (one body in Albion or unique individuals) or the nature of the Bible (priestly restrictions or poetic freedom), Blake seems to agree with it in one place and disagree with it in another. That kind of dialectical tension, which exists throughout *Jerusalem*, is most epigrammatically embodied in the phrase "Monos o Iesous." If we as readers can enter the paradoxes instead of fleeing them, we can find a way to transfigure the self-righteous, logical, sophisticated stones we want to throw, just as the resurrection of Jesus transforms the stony tomb.

As Jasper Hopkins explains the logic of Nicolas Cusanus:

> God's being is uniquely beyond all actual and conceptual differentiation, so that it cannot be truly and nonmetaphor-

> ically characterized by any predicate whose meaning is drawn from human experience.[25]

Like Cusanus, Blake leaps above ordinary human experience; that leap is embodied in the transfiguration. At the same time Blake plunges deeply into human experience: to deny the existence of adultery is to be non-human. By a leap of imagination, not only adultery but the whole law of Moses, and the whole law of Non-Contradiction, can be re-imagined. Thus a gesture that is completely strange to human experience—Jesus's answer to the scribes and Pharisees—turns out to embody the depths of human experience, where the human encounters and coincides with the divine. Pope's "To err is human, to forgive divine" receives its ultimate formulation when Blake combines the transfiguration and the story of the woman taken in adultery in his Divine Humanity.

In describing the dialectical method of Meister Eckhart, Cyprian Smith gives us a close analogy to Blake's method:

> in his preaching and writing, Eckhart keeps us perpetually swinging from one pole to the other; he will not let us rest in either. To rest in one and forget the other is to lose hold of the truth, which is essentially paradoxical. God is everything, yet nothing; distinct from creation, yet indistinct from it.... Having made a statement, Eckhart will often go on to deny it; but the truth lies neither in the affirmation nor in the denial, but in the tug-of-war between the two.[26]

Smith explains well the dynamic disequilibrium of such a dialectical method; add to that formula an absolute unity of total comprehension, just as sure as that of Alexander Pope, and you have Blake. According to Smith, the purpose of paradox in Eckhart and in Zen is to "bring the normal human intellect to the awareness of its own limitations, and thus open it up to the possibility of a higher kind of knowing."[27] While absolutely true of the learned ignorance of Cusanus and of Pope, this statement gives us only half of Blake. For in Cusanus and Pope we are to mistrust our own abilities, our own capacities for larger vision: we can never attain it, but can only acknowledge its existence. In Blake we fully participate in that higher Vision, as he takes the traditional formulae of the mystics as far as they can go:

> What is Eckhart's concept of Christ? It is of one in whom God and Man are made one. What is his concept of Man? It is of a being capable of becoming one with God. Eckhart is really only developing the ancient formula of Irenaeus and Athanasius: God became Man so that Man might become God.[28]

In the transfiguration, Jesus reveals that his physical body can become God and that therefore the spiritual God must have already become a physical body. In the story of the woman taken in adultery, Jesus reveals that the physical woman can not be separated from man nor from God. Any work of art is a disaster, a sin resulting from the fall, just like the sin of the woman taken in adultery. But at the same time Blake's art insists that this continuous falling away is, as in Christianity, the basis of forgiveness and therefore of resurrection and of transfiguration.

Jesus's confrontation with the scribes and Pharisees over the guilt and punishment of the woman taken in adultery reverses and collapses ordinary modes of being. The simultaneous contradictory meanings enact the resurrection paradox, which is contained in the motto "Monos o Iesous." The combination of forgiveness, which confronts and accepts sin, and transfiguration, which lifts God/Jesus in His most transcendent form above humanity, brings along with it not a loss of humanity, but a sacrificing of divinity—which the ultimate coincidences of opposites contained in Christianity express in the Incarnation and the Crucifixion. The dialectic of contradictions in words helps us get there, not only by encouraging us to seek beyond words, which is the vision of Percy Shelley and of Cusanus, but by embodying in words the very *coincidentia oppositorum* that is the basis of existence. The final vision is not bound by time or space; therefore language is inadequate to interpret it and can only enact it.

Albion's vision of Jesus on Plate 96 is a fulfillment of the transfiguration scene which was thwarted on Plate 4, where Albion refused to hear and see Jesus. Albion refused to see the combination of immanence and transcendence that the poet embraces.

As Jesus does in Luke with Moses and Elijah, so on *Jerusalem* 96 Jesus and Albion converse "as Man with Man" (96:6). While Albion is asking questions about the nature of self-sacrifice, the two are covered by a cloud: "the Covering Cherub coming on in darkness / Overshadowd them:" (96:17–18). The Covering Cherub is the equivalent here of the cloud which overshadows the transfiguration scene in Luke. And just as that cloud removes Jesus from the apostles, so here it removes Jesus, in the form of Los, from Albion: "the Cloud over-shadowing divided them asunder" (96:29). But unlike the apostles (Luke 9:34), who misunderstand the situation and fear for themselves:

> Albion stood in terror: not for himself but for his Friend
> Divine, & Self was lost in the contemplation of faith
> And wonder at the Divine Mercy. (96:30–32)

Albion has experienced the transfiguration, has awoken, realizes what he has been doing, and knows what now must be done.

> Do I sleep amidst danger to Friends! O my Cities & Counties
> Do you sleep! rouze up. rouze up. Eternal death is abroad
> ...
> Awake! Awake Jerusalem! O lovely Emanation of Albion
> Awake and overspread all Nations as in Ancient Time
> For lo! the Night of Death is past and the Eternal Day
> Appears upon our Hills: Awake Jerusalem, and come away.
> (96:33–34)

Having learned the lesson of self-annihilation (as in Jesus's birth), he stops trying to save himself by hiding in death, and gives himself up completely:

> ... Albion ... threw himself in the Furnaces of affliction
> All was a Vision, all a Dream: the Furnaces became
> Fountains of Living Waters flowing from the Humanity Divine.
> (96:35–37)

Just as forgiveness brought flowing waters out of Mary's agony, so sacrifice of Self leads to the joy of "living fountains of waters" (cf. Revelation 7:17), in a combination of forgiveness and transfiguration.

The final Vision has removed the doubt and despair which blocked such an awareness. A renewed humanity still creates space and time, but now they are created through imagination and not through fear and separation (98:30–31). Even death still exists:

> … & the all tremendous unfathomable Non Ens
> Of Death was seen in regenerations terrific or complacent varying
> According to the subject of discourse & every Word & Every Character
> Was Human according to the Expansion or Contraction, the Translucence or
> Opakeness of Nervous fibres such was the variation of Time & Space
> Which vary according as the Organs of Perception vary & they walked
> To & fro in Eternity as One Man reflecting each in each & clearly seen
> And seeing: according to fitness & order. (98:33–40)

This is exactly the same world which has appeared throughout *Jerusalem*, but now we see it from the perspective of eternal Vision instead of fallen vision. The Eye altering alters all. The opaqueness of Satan and the sensual contraction of Reuben still exist in this apocalypse, but when seen with eternal Vision they are shorn of their terrors.

The One Man in eternity has already spoken on Plate 4:

> I am not a God afar off, I am a brother and friend;
> Within your bosoms I reside, and you reside in me:
> Lo! we are One; forgiving all Evil; Not seeking recompense!
> Ye are my members O ye sleepers of Beulah, land of shades!
> (4:18–21)

Since they were spoken at the beginning of the poem, these echoes of Jeremiah 23:23, John 17:23, and John 15:4 do not suddenly appear at the end of *Jerusalem*; they could have been heard by Albion the first time. From this point of view, all the errors and terrors of *Jerusalem* have been simply a sleep. Like the apostles at the transfiguration, Albion needs but to awaken to see Jesus. But even though this fundamental safety net exists, the death and the agony are no less real, for man's fallen vision creates for him a life in which Jesus does not exist, for if Jesus were there, he would not die. Or, more exactly, with the presence of Jesus fully recognized, every death becomes a resurrection.

Transfiguration is a particularly apt word for Blake's art, even though he never used the word himself. For just as the copperplate artist exactly shapes his figures with engraving and etching, so Blake delineates his figures exactly, so that when they are reversed, they become completely other even while remaining essentially what they were.[29] In spiritual terms, the doubts and despairs are burned away so that all of existence stands clear. And the same existence that was so full of agony and separation is perceived to be eternity itself:

> All Human Forms identified even Tree Metal Earth & Stone. all
> Human Forms identified, living going forth & returning wearied
> Into the Planetary lives of Years Months Days & Hours reposing
> And then Awaking into his Bosom in the Life of Immortality.
> (99:1–4)

Even death itself, when perceived with eternal Vision, is reversed into larger life.

The transfiguration that Blake works on the Bible is based on a method of presenting oppositions and then transforming them, much as happens at the transfiguration of Jesus. "Monos o Iesous" combines apparently opposite qualities of Jesus which must be seen as simultaneous, even while mutually contradictory. In alluding to both these passages Blake puts us on notice that we are about to receive a new revelation. Blake thus puts himself in the line of and indeed in the place of the prophets and of Jesus himself in offering us a new voice which, addressed to Christian, Jew, and Deist, will offer us new ways to live as human souls.

Notes

1. Thomas J.J. Altizer, *The New Apocalypse* (Lansing: Michigan State UP, 1967), pp. 74–75.

2. *The Illuminated Blake*, p. 283.

3. Altizer, p. 73.

4. When the connection between Blake and Cusanus first occurred to me, I could find no one else who had made the connection. I was therefore delighted to see the 1980 publication of

Symbol and Truth in Blake's Myth, by Leopold Damrosch, who devotes several pages to the similarities. His most useful overall statement emphasizes how fundamental the *coincidentia oppositorum* is to both Blake and Cusanus:

> Like Blake, Cusanus stresses the *coincidentia oppositorum* and the supra-rational vision or *Nous* that apprehends their harmonious union in the Infinite. Now, the reconciliation of opposites was commonplace in the eighteenth century; as E.R. Wasserman has shown, *concordia discors* is the fundamental principle of Denham's *Cooper Hill* and Pope's *Windsor Forest....* But these poets postulated a God above and beyond the sublunary realm who harmonized opposites by manipulating them in a larger whole, the lights and shadows of a grand design. Cusanus, like Blake in his 'marriage' of heaven and hell, sought a vital *union* of opposites in the ultimate order of things, not a reconciliation controlled, as it were, from above. For Cusanus the ultimate order is God, 'because He is Himself the Absolute Ground, in which all otherness (*alteritas*) is unity, and all diversity is identity.' By the time of the prophetic books, Blake too invested Jesus, in however heterodox a manner, with the same status of ultimate ground of diversity-in-identity. (p. 21)

5. Michael Tolley, "William Blake's Use of the Bible" (Dissertation, University of London, 1974), p. 367.

6. *The Poems of William Blake* (NY: Longman, 1989), p. 630.

7. Ostriker, *William Blake,* p. 997.

8. Joseph Wicksteed, *William Blake's Jerusalem: A Commentary* (London: Trianon, n.d.), pp. 117–18.

9. Anne Mellor, *Blake's Human Form Divine* (Berkeley: Univ. of California Press, 1974), pp. 292–93.

10. *Blake's Poetry and Designs* (NY: Norton, 1979), p. 313.

11. 30 January 1803, E727.

12. Behrendt, *Reading,* p. 165.

13. The same story is told in Matthew 17 and Mark 9, without the word "Monos" next to "Iesous" and with the significant omission of the apostles' falling asleep.

14. *Prophecy and the Philosophy of Mind: Traditions of Blake and Shelley* (Univ. of Alabama, 1985), p. 74.

15. Hoagwood, p. 78.

16. *The Illuminated Blake*, p. 284.

17. Compare this observation by W.J.T. Mitchell:

> his choices of unusual subjects (Newton inscribing his mathematical diagrams, ... Christ writing on the ground to confound the scribes and pharisees) suggest that the moment of inscription tended to stand out for him as a principal subject for illustration in any narrative. ("Visible," p. 64)

18. Carl Jung, *Answer to Job*, trans. R.F.C. Hull (1958; rpt. Princeton, 1973), p. 81.

19. Compare this stunning and hilarious outburst from the Devil in *The Marriage of Heaven and Hell*:

> if Jesus Christ is the greatest man, you ought to love him in the greatest degree; now hear how he has given his sanction to the law of ten commandments: did he not mock at the sabbath ... [other examples] I tell you, no virtue can exist without breaking these ten commandments: Jesus was all virtue, and acted from impulse: not from rules. (E43)

20. *Jerusalem* 37, mirror writing in illumination.

21. *The Divine Image* (Rome: Edizioni di Storia e Letteratura, 1950), p. 9.

22. Altizer, p. xi.

23. I do not subscribe to the theories of the "anxiety of influence" as proposed by Harold Bloom in, for example, *The Anxiety of Influence* (NY: OUP, 1973). Joseph Anthony Wittreich, Jr., writing specifically of Blake's relationship to Milton, counters Bloom's ideas:

> The notion that Blake suffered under the anxiety of influence, that he was obsessed by a drive to transcend Milton, even at the cost of misrepresenting and misinterpreting him, is antithetical not only to the spirit of Blake's rejoinder to Lavater but to the principle of art that Blake enunciates in *A Descriptive Catalogue*: 'To suppose that Art can go beyond the finest specimens of Art that are now in the world,' Blake says, 'is not knowing what Art is; it is being

> blind to the gifts of the spirit' (p. 535 [i.e., E535]). (*Angel of Apocalypse*, p. 250)

Furthermore, Bloom's theory is based on a metaphor of limited space: the strong poet must shoulder aside his predecessor because there is not enough space for both of them to occupy. Blake, however, does not assume a finite space which only one poet can occupy at a time. Instead, each artist-Christian, in the expression of his own Poetic Genius, is at once a unique individual and a manifestation of Eternity. The question of fighting for elbow room simply does not arise. The inextricability of fallen and eternal Vision is not a concession to fallen vision and its shrunken notions of space and time; it is rather a transfiguration of space and time into a larger understanding.

24. In her dissertation, "William Blake's New Typology and the Revaluation of Prophecy in the Eighteenth Century" (Emory University, 1979), Patricia Elizabeth Davis studies *Europe*, *Milton*, and the Job illustrations in the context of eighteenth-century interpretations of the Bible. According to Davis, "Blake invented a new typology, transforming his prophetic poems into primers for creative exegesis" (p. 2). Basing her study on an idea of *recta ratio*, "which affirms that what a man knows depends upon what, as a moral being, he chooses to make himself" (p. 3), she compares Blake's progressive revelation through prophecy to the methods of his contemporaries, such as Calvin and Wesley.

Davis's dissertation is part of the effort to rectify "the paucity of background studies necessary for a pursuit of Blake's relationship to the biblical traditions available in his own time," which is lamented by Leslie Tannenbaum in *Biblical Tradition in Blake's Early Prophecies* (Princeton UP, 1982), p. x. Tannenbaum's book raises typology to a definition of Blake's art: "By defining Christ as the Imagination, Blake adopts as his subject the typological process itself" (p. 99).

Setting Blake's early prophecies within their contemporary biblical traditions, Tannenbaum sees them, especially *The Book of Los*, as containing "a prospect of the road leading to *Jerusalem*" (280) because they invert and parody but do nothing direct to "bring to a period the cyclical pattern" (p. 280) which the Bible would be without the Book of Revelation:

> Through Los's forging of the sun in *Asia* and in *The Book of Los*, Blake reveals that he has begun to realize that the road to Eternity is the road that leads ahead, leading the Imagination further into the demon universe; it must continue to create the human illusion until this illusion has been embodied in all its variety at the same time that it becomes condensed into a single unifying image. Only then, as in Revelation, when the body of error has been completely revealed, will the body of truth reappear, will the heavenly bride descend and meet the bridegroom, and will Blake simultaneously fulfil and annihilate the great code of art. (p. 281)

I am pleased to see that this final paragraph of Tannenbaum's book confirms my approach here.

25. *Nicholas of Cusa's Dialectical Mysticism: Text, Translation, and Interpretive Study of De Visione Dei*. 2nd rev ed. (Minneapolis: The Arthur J. Banning Press, 1988), p. 41.

26. Cyprian Smith, *The Way of Paradox* (London: Darton, Longman and Todd, 1987), p. 27.

27. Cyprian Smith, p. 27.

28. Cyprian Smith, p. 77.

29. For a fuller discussion of the reversals in Blake's physical act of producing his illuminated books, see Chapter 7.

Conclusion
Into the Fourth Dimension

The solemn tone and religious emphasis of my exploration into the *coincidentia oppositorum* conceal the basic humor that springs from such an affront to common sense. Like most jokes and riddles, the *coincidentia oppositorum* both fulfills and thwarts expectations. The burst of laughter that follows a good joke erupts from the coming together of two opposing forces in the punch line: the listener is surprised at the ending but simultaneously pleased by its appropriateness.

Two eighteenth-century gentlemen observe a coach with three passengers inside. "Behold," says the first gentleman, "an emblem of the Trinity: three persons in one vehicle." "Not so," responds his companion. "If you want to see an emblem of the Trinity, you must see one person in three vehicles."[1] Anyone who finds this joke funny is struck first with the surprise; like Samuel Johnson responding to the metaphysical poets, she may wonder how anyone could ever think of something so unusual. At the same time, the laughing listener must admit that the punch line follows inevitably from the set-up. This combination of surprise and fulfillment of expectation is exactly the pattern of the *coincidentia oppositorum*.

Every knock-knock joke uses a pun to follow this same structure. "Knock, knock." "Who's there?" "Orange." "Orange who?" "Orange you glad to see me?" Children especially delight in the potential of language and the sounds of language to surprise and to fulfill expectations at the same time. The *coincidentia oppositorum* is the grand cosmic laughter of our universe, both absurdly impossible and logically coherent. Living under the sway of ideologies that tell us what is natural, we seldom question the status quo. And as we see so often, in the French Revolution and in *Animal Farm*, apparent rebels who do seem to challenge the powers that be often reveal themselves to be tyrants as bad as the ones they overthrew.

The Law of Non-Contradiction insists that we must live an orderly, respectable spiritual and mental life. Cusanus's vision of God and Blake's vision of the divine humanity laugh in the face of such consistent respectability, but they are not at all nihilistic, for even while mocking, they build an alternative way of existing that bases itself on freedom and creativity. The laughter of the artist, the smile of the saint, the sparkle of the mischief maker infects us when we read them.

The idea of the *coincidentia oppositorum* is a very simple geometrical, logical one. We can have it both ways: the square can be a circle, the triangle can be a line; good can be evil, a sin can be a blessing, a blessing can be a sin, God can be man, man can be God. And even if a reader of Blake is not herself religious, she can still admit the simple fact of life that impossibilities and improbabilities happen every moment, around us and in us.

As I prepare to send my manuscript to the publisher, the cover of *The New Republic* (12 July 1993) features "A Plea for Higher Dimensions." Jim Holt's article, "The Newer Paradigm" (pp. 22–25) whimsically wonders how we might pass beyond Aristotle's limitation that "the three dimensions are all that there are." Sounding remarkably like Blake, Holt explains how mathematicians produced a "triumph of the imagination over the senses" and "an antidote to the evils of materialism" when they demonstrated "the mathematical reality of higher dimensions."

"How do you know but ev'ry Bird that cuts the airy way, / Is an immense world of delight, clos'd by your senses five?" asks Blake in *The Marriage of Heaven and Hell*. "Might we not be surrounded by wondrous caverns of possibility that our eyes do not yet behold?" muses Holt. Evoking the world of E.A. Abbott's 1884 *Flatland: A Romance in Many Dimensions*, in which two-dimensional beings have trouble imagining a third-dimension, Holt uses geometry, as does Cusanus, to try to raise our imaginations beyond its current limits. And as does Blake in the quotation from *Jerusalem* 5 near the beginning of my Chapter 7, so Holt suggests that the way out is the way in:

> Maneuvering oneself into the fourth dimension means moving, in a rather special sense, toward one's inside, in the direction of one's heart.... The fourth dimension will permit all contradictions to be resolved.[2]

Hazard Adams declares that he reads Blake to feel "the dizziness of freedom."[3] Part of that dizziness certainly must arise from the raucous disturbance caused by the cosmic guffaw of Blake's *coincidentia oppositorum,* which demands that we simultaneously lose our equilibrium and stand firm.

Notes

1. I consider this joke the best one told by Leo Damrosch during his 1991 NEH Summer Seminar for college teachers.

2. Holt, p. 24.

3. "The Dizziness of Freedom," *College English,* 48 (1986), pp. 431–43.

Works Consulted

Abrams, M.H. *The Mirror and the Lamp: Romantic Theory and the Critical Tradition.* 1953; rpt. London: Oxford UP, 1976.

———. *Natural Supernaturalism: Tradition and Revolution in Romantic Literature.* New York: Norton, 1971.

Adams, Hazard. *Critical Essays on William Blake.* Boston: G.K. Hall, 1991.

———. "The Dizziness of Freedom; or, Why I Read William Blake." *College English* 48, no. 5 (Sept 1986): 431–43.

———. "The Mental Traveller." In *William Blake: A Reading of the Shorter Poems.* Seattle: Univ. of Wash, 1963. 77–100

———. *Philosophy of the Literary Symbolic.* Tallahassee: University Presses of Florida, 1983.

———. "Synecdoche and Method." *Critical Paths* 41–71.

Adamson, Arthur. "Structure and Meaning in Blake's 'The Mental Traveller.'" *Mosaic* 7 (1974): 41–58.

Altizer, Thomas J.J. *The New Apocalypse: The Radical Christian Vision of William Blake.* Lansing: Michigan State UP, 1967.

Aristotle. *Metaphysica.* Trans. W.D. Ross. *The Works of Aristotle Translated into English.* Vol. 8: *Metaphysica.* 2nd ed. Oxford, 1928.

———. *The Metaphysics.* Trans. Hugh Tredennick. Loeb Classical Library. Cambridge: Harvard UP, 1936.

Auerbach, Erich. *Mimesis: The Representation of Reality in Western Literature*. Trans. Willard R. Trask. Princeton: Princeton UP, 1968.

Ault, Donald D. *Visionary Physics: Blake's Response to Newton*. Chicago: University of Chicago Press, 1974.

Barfield, Owen. *What Coleridge Thought*. Middletown, CT: Wesleyan UP, 1971.

Barth, J. Robert. *Coleridge and Christian Doctrine*. Cambridge: Harvard UP, 1969.

———. *Coleridge and the Power of Love*. Columbia: Univ. of Missouri Press, 1988.

Bateman, Arthur B. "The Spiritual Genius of William Blake." *London Quarterly Review* 148 (1927): 34–45.

Beer, John. *Blake's Visionary Universe*. New York: Barnes and Noble, 1969.

Behrendt, Stephen, ed. *Approaches to Teaching Shelley's* Frankenstein. NY: MLA, 1990.

———, ed. *History & Myth: Essays on English Romantic Literature*. Detroit: Wayne State UP, 1990.

———. *Reading William Blake*. NY: St Martin's Press, 1992.

Bennett, Betty T. "Love and Egocentricity: Teaching *Alastor* and *Prometheus Unbound* with Mary Shelley's *Frankenstein*." In Hall, *Approaches*. 76–78.

Bentley, G.E., Jr. *Blake Records*. Oxford: Clarendon Press, 1969.

———. *Blake Records Supplement: Being New Materials Relating to the Life of William Blake Discovered Since the Publication of Blake Records (1969)*. Oxford: Clarendon, 1988.

Berger, P. *William Blake: Poet and Mystic*. Trans. Daniel H. Cormer. London: Chapman and Hall, 1914.

Berger, Peter. "The Other Side of God—Problem and Agenda." In Berger, *The Other Side*. 3–27.

———, ed. *The Other Side of God: A Polarity in World Religions*. NY: Anchor/Doubleday, 1981.

Bett, Henry. *Nicholas of Cusa.* London: Methuen, 1932.

Biechler, James E. *The Religious Language of Nicholas of Cusa.* Missoula, MT: Scholars Press, 1975.

Billigheimer, Rachel V. *Wheels of Eternity: A Comparative Study of William Blake and William Butler Yeats.* NY: St. Martin's, 1990.

Birkby, Dana Bliss. *Salting the Tail of the Infinite: Mathematics in the Epistemology of Nicholas of Cusa.* Unpublished Thesis. Harvard, 1982.

Blackstone, Bernard. *English Blake.* 1949; rpt. Hamden, CT: Shoestring Press, 1966.

Blake, William. *Blake's Poetry and Designs.* Eds. Mary Lynn Johnson and John E. Grant. NY: Norton, 1979.

———. *The Complete Poetry and Prose of William Blake.* Ed. David V. Erdman. Commentary by Harold Bloom. Newly revised ed. Garden City, NY: Anchor/Doubleday, 1982.

———. *The Illuminated Blake.* Ed. David V. Erdman. Garden City, NY: Anchor/Doubleday, 1974.

———. *Poems of William Blake.* Ed. William Butler Yeats. 1905; rpt. London: Routledge and Kegan Paul, 1969.

———. *The Poems of William Blake.* 2nd ed. Ed. W.H. Stevenson. NY: Longman, 1989.

———. *William Blake: Jerusalem: The Emanation of the Giant Albion.* Blake's Illuminated Books, Vol. 1. Ed. Morton Paley. William Blake Trust/Princeton UP, 1991.

Bloom, Harold. *The Anxiety of Influence: A Theory of Poetry.* NY: Oxford UP, 1973.

———. *Blake's Apocalypse.* Ithaca: Cornell UP, 1963.

———, ed. *Deconstruction and Criticism.* NY: Seabury Press, 1979.

———. "Introduction." In Bloom, *Mary Shelley's Frankenstein.* 1–11.

———. *Mary Shelley's Frankenstein: Modern Critical Interpretations.* NY: Chelsea House, 1987.

———. *Poetry and Repression: Revisionism from Blake to Stevens*. New Haven: Yale UP, 1976.

———. *Shelley's Mythmaking*. Ithaca: Cornell UP, 1959.

Blunt, Anthony. "Blake's 'Ancient of Days': The Symbolism of the Compasses." In *The Visionary Hand: Essays for the Study of William Blake's Art and Aesthetics*. Ed. Robert Essick. Los Angeles: Hennessey and Ingalls, 1973. 71–103.

Bostetter, Edward E. *The Romantic Ventriloquists: Wordsworth, Coleridge, Keats, Shelley, Byron*. Revised edition. Seattle: Univ. of Washington, 1975.

Botting, Fred. *Making Monstrous:* Frankenstein, *Criticism, Theory*. Manchester and NY: Manchester UP and St. Martin's Press, 1991.

Bottrall, Margaret. *The Divine Image: A Study of Blake's Interpretation of Christianity*. Rome: Edizioni di Storia e Letteratura, 1950.

Bouwer, Izak, and Paul McNally. "'The Mental Traveller': Man's Eternal Journey." *Blake: An Illustrated Newsletter* (Winter 1978–79): 184–92.

Bracher, Mark. *Being Form'd: Reading Through Blake's* Milton. Station Hill Press, 1985.

Brisman, Leslie. *Milton's Poetry of Choice and Its Romantic Heirs*. Ithaca: Cornell UP, 1973.

———. *Romantic Origins*. Ithaca: Cornell UP, 1978.

Brooks, Harold F. "Blake and Jung II: The Fall into Disintegration." *Aligarh Critical Miscellany* 2 (1989): 158–74.

Brooks, Peter. "'Godlike Science/Unhallowed Arts': Language, Nature, and Monstrosity." In Levine and Knoepflmacher, *Endurance*. 205–20.

Brown, Norman O. *Life Against Death*. New York: Modern Library, 1959.

Brown, Raymond E. *The Birth of the Messiah: A Commentary on the Infancy Narratives in Matthew and Luke*. NY: Doubleday, 1977.

Bultmann, Rudolf. *History and Eschatology: The Presence of Eternity*. NY: Harper, 1957.

———. *Theology of the New Testament*. 2 vols. Trans. Kendrick Grobel. NY: Scribners, 1951, 1955.

Byron, George Gordon. *The Poetical Works of Byron*. Ed. Robert F. Gleckner. Boston: Houghton Mifflin, 1975.

———. *Byron's Poetry*. Ed. Frank McConnell. NY: Norton, 1978.

Canfield, Douglas. "The Fate of the Fall in Pope's *Essay on Man*." *The Eighteenth Century* 23 (1982): 134–50.

Cantor, Paul A., and Michael Valdez Moses. "Teaching *Frankenstein* from the Creature's Perspective." In Behrendt, *Approaches*. 127–32.

Cassirer, Ernst. *The Individual and the Cosmos in Renaissance Philosophy*. Trans. Mario Domandi. New York: Harper, 1963.

Chayes, Irene. "Plato's *Statesman* Myth in Shelley and Blake." *Comparative Literature* 13 (1961): 358–69.

Chernaik, Judith. *The Lyrics of Shelley*. Cleveland: Case Western Reserve U, 1972.

Cherniss, Harold. *Aristotle's Criticism of Presocratic Philosophy*. NY: Octagon, 1983.

Clarke, John Henry. *The God of Shelley and Blake*. 1929; rpt. London: Haskell House, 1966.

———. *William Blake on the Lord's Prayer*. 1927; rpt. London: Haskell House, 1971.

Cody, Richard. *The Landscape of the Mind*. Oxford: Oxford UP, 1969.

Coleman, Deirdre. *Coleridge and The Friend (1809–1810)*. Oxford: Clarendon, 1988.

Coleman, Deirdre, and Peter Otto, eds. *Imagining Romanticism: Essays on English and Australian Romanticisms*. West Cornwall, CT: Locust Hill Press, 1992.

Coleman, Deirdre, and Peter Otto. "Introduction." In Coleman and Otto. ix–xxi.

Coleridge, Samuel Taylor. *Biographia Literaria*. Ed James Engell and W. Jackson Bate. Bollingen Series LXXV. Princeton UP, 1983.

———. *Biographia Literaria: or Biographical Sketches of My Literary Life and Opinions*. Ed. George Watson. Dent: Everyman's Library, 1982.

———. *Collected Letters of Samuel Taylor Coleridge*. 6 vols. Ed. Earl Leslie Griggs. Oxford, 1956–71.

———. *Notebooks*. 3 vols. Ed. Kathleen Coburn. Princeton UP, 1957–73.

———. *On the Constitution of Church and State According to the Idea of Each*. Ed. John Barrell. London: Dent, 1972.

———. *The Portable Coleridge*. Ed. I.A. Richards. NY: Viking, 1950.

———. *Poetical Works*. Ed. Ernest Hartley Coleridge. 1912; rpt. Oxford UP, 1978.

Conger, Syndy M. "Aporia and Radical Empathy." In Behrendt, *Approaches*. 60–66.

Cooke, Michael G. *Acts of Inclusion: Studies Bearing on an Elementary Theory of Romanticism*. New Haven: Yale UP, 1979.

———. *The Romantic Will*. New Haven: Yale UP, 1976.

Copi, Irving M. *Introduction to Logic*. Seventh Edition. NY: Macmillan, 1986.

Copleston, Frederick. *Thomas Aquinas*. London: Search Press, 1976.

Cousins, Ewert H. "Francis of Assisi and Bonaventure: Mysticism and Theological Interpretation." In Berger, *The Other Side*.

Cues, Nicolas de [Nicholas of Cusa]. *Lettres aux moines de Tegernsee sur la docte ignorance (1452–1456), Du Jeu de la Boule (1463)*. Ed. Maurice de Gandillac. Paris: O.E.I.L., 1985.

Curran, Stuart. "The Siege of Hateful Contraries." In *Milton and the Line of Vision*. Ed. Joseph Anthony Wittreich, Jr. Madison: University of Wisconsin Press, n.d.

———. "The Structures of *Jerusalem.*" In *Blake's Sublime Allegory.* Ed. Stuart Curran and Joseph Anthony Wittreich, Jr. Madison: University of Wisconsin Press, 1973. 329–46.

Cusanus, Nicolas. *Of Learned Ignorance.* Trans. Germain Heron. Introduction by J.B. Hawkins. New Haven: Yale UP, 1954.

Damon, S. Foster. *A Blake Dictionary: The Ideas and Symbols of William Blake.* Ed. Morris Eaves. Boulder, CO: Shambhala, 1979.

———. *William Blake: His Philosophy and Symbols.* Boston: HM, 1924.

Damrosch, Leopold, Jr. *Symbol and Truth in Blake's Myth.* Princeton UP, 1980.

Daniel-Rops, Henri. *The Church in an Age of Revolution: 1789–1870.* Trans. John Warrington. NY: Dutton, 1965.

Davies, J.G. *The Theology of William Blake.* 1948; rpt. Archon Books, 1966.

Davis, Patricia Elizabeth. "William Blake's New Typology and the Revaluation of Prophecy in the Eighteenth Century." Diss. Emory University, 1979.

Dawson, P.M.S. "'The Mask of Darkness': Metaphor, Myth, and History in Shelley's 'The Triumph of Life.'" In Behrendt, *History.* 235–44.

de Man, Paul. "Shelley Disfigured." In Bloom, *Deconstruction.* 39–73.

Dolan, John Patrick. *Unity and Reform: Selected Writings of Nicholas de Cusa.* Univ. of Notre Dame Press, 1962.

Donoghue, Denis. "The Flight of Gerard Manley Hopkins." Review of *Gerard Manley Hopkins: A Very Private Life* by Robert Bernard Martin (Putnam, 1991). *New York Review of Books* (18 July 1991): 14–18.

Dorfman, Deborah. "Knowledge and Estimation of Blake during His Lifetime." In Adams, *Critical Essays.* 11–21.

Doskow, Minna. *William Blake's* Jerusalem: *Structure and Meaning in Poetry and Picture.* Rutherford, NJ: Fairleigh Dickinson UP, 1982.

Dunn, Richard J. "Narrative Distance in *Frankenstein*." *Studies in the Novel* (North Texas State University) 6 (1974): 408–17.

Easson, Roger. "William Blake and His Reader in *Jerusalem*." In *Blake's Sublime Allegory*. Ed. Stuart Curran and Joseph Anthony Wittreich, Jr. Madison: University of Wisconsin Press, 1973.

Eaves, Morris, and Michael Fischer, eds. *Romanticism and Contemporary Criticism*. Ithaca: Cornell UP, 1986.

Eckhart, Meister. *The Essential Sermons, Commentaries, Treatises, and Defense*. Translation and Introduction by Edmund Colledge, O.S.A. and Bernard McGinn. Preface by Huston Smith. NY: Paulist Press, 1981.

Eco, Umberto. *The Aesthetics of Thomas Aquinas*. Trans. Hugh Bredin. Cambridge: Harvard UP, 1988.

Efird, James M., ed. *The Use of the Old Testament in the New and Other Essays: Studies in Honor of William Franklin Stinespring*. Durham: Duke UP, 1972.

Eliade, Mircea. *The Sacred and the Profane: The Nature of Religion*. Trans. Willard Trask. NY: Harcourt, 1959.

Ellis, Edwin J. *The Real Blake: A Portrait Biography*. NY: McClure, Phillips and Co., 1907.

———, and William Butler Yeats. *The Works of William Blake*. 3 vols. 1893; rpt. NY: AMS Press, 1973.

Enscoe, Gerald E. "The Content of Vision: Blake's 'Mental Traveller.'" *Papers on Language & Literature*, 4 (1968): 400–13.

Erdman, David V., ed. *Blake and His Bibles*. West Cornwall, CT: Locust Hill Press, 1990.

———. *Blake: Prophet Against Empire*. 3rd ed. Princeton: Princeton UP, 1977.

———, et al., eds. *A Concordance to the Writings of William Blake*. 2 vols. Ithaca: Cornell UP, 1967.

Essick, Robert N. *William Blake and the Language of Adam*. Oxford: Clarendon, 1989.

———. *William Blake, Printmaker*. Princeton: Princeton UP, 1980.

Favret, Mary A. *Romantic Correspondence: Women, Politics, and the Fiction of Letters*. Cambridge: Cambridge UP, 1993.

Ferguson, James. "Prefaces to *Jerusalem*." In *Interpreting Blake*. Ed. Michael Phillips. Cambridge: Cambridge UP, 1979. 164–95.

Ferguson, John. *Aristotle*. NY: Twayne, 1972.

Fisch, Harold. *Jerusalem and Albion: The Hebraic Factor in Seventeenth-Century Literature*. NY: Schocken, 1964.

Fish, Stanley E. *Self-Consuming Artifacts: The Experience of Seventeenth-Century Literature*. Berkeley: University of California Press, 1972.

Fishbane, Michael. "Israel and the 'Mothers.'" In Berger, *The Other Side*. 28–47.

Fisher, Peter F. *The Valley of Vision: Blake as Prophet and Revolutionary*. Ed. Northrop Frye. Toronto: University of Toronto Press, 1961.

Fogle, Richard Harter. "A Note on Romantic Oppositions and Reconciliations." *The Major English Romantic Poets: A Symposium in Reappraisal*. Ed. Clarence D. Thorpe, et al. Carbondale: Southern Illinois UP, 1957. 17–23.

Frei, Hans W. *The Eclipse of Biblical Narrative: A Study in Eighteenth and Nineteenth Century Hermeneutics*. New Haven: Yale University Press, 1974.

Frosch, Thomas R. *The Awakening of Albion: The Renovation of the Body in the Poetry of William Blake*. Ithaca: Cornell UP, 1974.

Frye, Northrop. "The Archetypes of Literature." In *Fables of Identity: Studies in Poetic Mythology*. New York: Harcourt, 1963.

———. "Blake's Reading of the Book of Job." In *William Blake: Essays for S. Foster Damon*. Ed. Rosenfeld. 221–34.

———. *The Double Vision: Language and Meaning in Religion*. Univ of Toronto Press, 1991.

———. *Fearful Symmetry: A Study of William Blake*. Princeton: Princeton UP, 1969.

———. *The Great Code: The Bible and Literature*. NY: Harcourt, 1982.

Furst, Lilian R. *Romanticism*. London: Methuen, 1969.

Gallant, Christine. *Blake and the Assimilation of Chaos*. Princeton: Princeton UP, 1978.

Gelpi, Barbara Charlesworth. *Shelley's Goddess: Maternity, Language, Subjectivity*. NY: Oxford UP, 1992.

George, Diana Hume. *Blake and Freud*. Ithaca: Cornell UP, 1980.

Gilby, Thomas. *Poetic Experience: An Introduction to Thomist Aesthetic*. London: Sheed & Ward, 1934.

Gilchrist, Alexander. *A Life of William Blake*. Vol. 1. 1880; rpt. Totowa, NJ: Rowman and Littlefield, 1973. 2 vols.

Gillet, Louis. "Un Mystique Anglais: William Blake." *Revue des Deux Mondes* 38 (1937): 190–206.

Gilson, Etienne. *The Philosophy of St. Thomas Aquinas*. Authorised Translation from the Third Revised and Enlarged Edition of 'Le Thomisme.' Trans. Edward Bullough. Salem, NH: Ayer Co., 1983.

Goddard, Harold C. *Blake's Fourfold Vision*. Wallingford, PA: Pendle Hill, 1956.

Goldgar, Bertrand. "Pope's Theory of the Passions: The Background of Epistle II of the *Essay on Man*." *Philological Quarterly* 41 (1962): 730–43.

Goldsmith, N.M. "A Reconciliation of Opposites: Concepts of the Mind in Pope and Coleridge." *Prose Studies* 7 (1984): 3–25.

Grabo, Carl. *The Magic Plant: The Growth of Shelley's Thought*. Chapel Hill: Univ. of NC Press, 1936.

Grimes, Ronald L. *The Divine Imagination: William Blake's Major Prophetic Visions*. Metuchen, NJ: The Scarecrow Press and the American Theological Library Association, 1972.

Hall, Spencer, ed. *Approaches to Teaching Shelley's Poetry*. NY: MLA, 1990.

Harding, Anthony John. *Coleridge and the Inspired Word*. Montreal: McGill-Queen's UP, 1985.

Harper, George Mills. "The Divine Tetrad in Blake's *Jerusalem*." In Rosenfeld, *William Blake*. 235–55.

Hawkins, D. J. B. "Introduction." In Cusanus, *Learned*.

Helms, Randel. "Blake's Use of the Bible in 'A Song of Liberty.'" *English Language Notes* 16 (1979): 287–91.

———. "Ezekiel and Blake's *Jerusalem*." *Studies in Romanticism* 13 (1974): 127–40.

Herson, Ellen Brown. "Oxymoron and Dante's Gates of Hell in Shelley's *Prometheus Unbound*." *Studies in Romanticism* 29 (1990): 371–93.

Heurtevent, Raoul. *Durand de Troarn et les Origines de l'Hérésie Berengarienne*. Paris: Beauchesne, 1912.

Hildebrand, W.H. "Shelley's Early Vision Poems." *Studies in Romanticism* 8 (1969): 216–28.

Hoagwood, Terence Allan. *Prophecy and the Philosophy of Mind: Traditions of Blake and Shelley*. Univ. of Alabama, 1985.

Hodgson, John A. *Coleridge, Shelley, and Transcendental Inquiry: Rhetoric, Argument, Metapsychology*. Lincoln: Univ. of Nebraska Press, 1989.

Holmes, Richard. *Coleridge*. Past Masters Series. OUP, 1982.

———. *Shelley: The Pursuit*. NY: Dutton, 1975.

Holt, Jim. "The Newer Paradigm: A Plea for Higher Dimensions." *The New Republic*, July 12, 1993, pp. 22–25.

Hopkins, Jasper. *A Concise Introduction to the Philosophy of Nicholas of Cusa*. 3rd edition. Minneapolis: The Arthur J. Banning Press, 1986.

———. *Nicholas of Cusa on Learned Ignorance: A Translation and an Appraisal of De Docta Ignorantia*. Minneapolis: The Arthur J. Banning Press, 1981.

———. *Nicholas of Cusa's Debate with John Wenck: A Translation and an Appraisal of De Ignota Litteratura and Apologia*

Doctae Ignorantiae. Minneapolis: The Arthur J. Banning Press, 1981.

———. *Nicholas of Cusa's Dialectical Mysticism: Text, Translation, and Interpretive Study of De Visione Dei.* 2d rev. ed. Minneapolis: The Arthur J. Banning Press, 1988.

Howard, John. *Infernal Poetics: Poetic Structures in Blake's Lambeth Prophecies.* Rutherford: Fairleigh Dickinson Univ. Press, 1984.

Hughes, William R., ed. *Jerusalem: A Simplified Version.* London: George Allen and Unwin, 1964.

Humphreys, Christmas. *Buddhism.* 1951; rpt. NY: Penguin, 1975.

The Interpreter's Bible. Ed. Norman H. Snaith and Ralph W. Sockman. Nashville: Abingdon, 1954.

The Interpreter's Bible. Vol. III. Ed. Norman H. Snaith and Ralph W. Sockman. Nashville: Abingdon, 1954.

The Interpreter's Bible. Vol. VIII. Ed. S. MacLean Gilmorn and Wilbert F. Howard. Nashville: Abingdon, 1952.

The Interpreter's Dictionary of the Bible: An Illustrated Encyclopedia. 4 vols. Nashville: Abingdon, 1962.

Ivanov, V.V. "The Semiotic Theory of Carnival as the Inversion of Bipolar Opposites." *Carnival!* Ed. Thomas A. Sebeok. New York: Mouton, 1984. 11–35.

Jasper, David. *Coleridge as Poet and Religious Thinker: Inspiration and Revelation.* London: Macmillan, 1985.

Jenkins, Patricia Mavis. *Coleridge's Literary Theory: The Chronology of Its Development, 1790–1818.* Fairfield, CT: Fairfield U, 1984.

Joseph, M.K. "Introduction." In Shelley, *Frankenstein.* OUP.

Jung, Carl Gustave. Trans. R.F.C. Hull. *Aion: Researches into the Phenomenology of the Self.* 2nd edition. 1959; rpt. Princeton UP, 1979.

———. *Answer to Job.* Trans. R.F.C. Hull. 1958; rpt. Princeton UP, 1973.

———. *Modern Man in Search of a Soul.* Trans. W.S. Dell and Cary F. Baynes. NY: Harcourt, 1933.

Kabitoglou, E. Douka. *Plato and the English Romantics: Dialogoi.* NY: Routledge, 1990.

Kallich, Martin. *Heav'n's First Law.* DeKalb, IL: Northern Illinois UP, 1967.

———. "Unity and Dialectic: The Structural Role of Antitheses in Pope's *Essay on Man.*" *Papers on Language and Literature* 1 (1965): 109–24.

Keach, William. *Shelley's Style.* New York: Methuen, 1984.

Kestner, Joseph. "Narcissism as Symptom and Structure: The Case of Mary Shelley's *Frankenstein.*" In *The Nature of Identity: Essays Presented to Donald E. Hayden by the Graduate Faculty of Modern Letters.* Tulsa: University of Tulsa, 1981. 15–25.

Keynes, Geoffrey, ed. *Blake's Illustrations to the Bible.* London: Trianon Press, 1956.

———, ed. *The Letters of William Blake.* London: Rupert Hart-Davis, 1968.

———, ed. *William Blake: Complete Writings.* London: Oxford UP, 1972.

Kiely, Robert. *The Romantic Novel in England.* Cambridge: Harvard UP, 1972.

Kiralis, Karl. "A Guide to the Intellectual Symbolism of William Blake's Later Prophetic Writings." *Criticism* 1 (1959): 190–210.

———. "Theme and Structure of Blake's *Jerusalem.*" *English Literary History* 23 (1956): 127–43.

Klibansky, Raymond. "Plato's *Parmenides* in the Middle Ages and the Renaissance: A Chapter in the History of Platonic Studies." *Medieval and Renaissance Studies* 1 (1943): 281–330.

Knapp, Elise F. "Palladio, Pope, and the Design of *An Essay on Man.*" *Essays in Arts and Sciences* 19 (1990): 65–81.

Knight, G. Wilson. *Laureate of Peace: On the Genius of Alexander Pope*. New York: Oxford University Press, 1955.

Korteling, Jacomina. *Mysticism in Blake and Wordsworth*. NY: Haskell House, 1966.

Koyre, Alexandre. *From the Closed World to the Infinite Universe*. New York: Harper, 1957.

Kroeber, Karl. "Delivering *Jerusalem*." *Blake's Sublime Allegory*. Ed. Stuart Curran and Joseph Anthony Wittreich, Jr. Madison: University of Wisconsin, 1973.

Lawlor, Nancy. "Pope's *Essay on Man*: Oblique Light for a False Mirror." *Modern Language Quarterly* 28 (1967): 305–16.

Leask, Nigel. *The Politics of Imagination in Coleridge's Critical Thought*. NY: St. Martin's, 1988.

Levao, Ronald. *Three Renaissance Minds: Cusanus, Sidney, Shakespeare*. Berkeley: Univ. of CA Press, 1985.

Levine, George, and U.C. Knoepflmacher, eds. *The Endurance of* Frankenstein: *Essays on Mary Shelley's Novel*. Berkeley: Univ. of California Press, 1979.

Lovejoy, Arthur. *The Great Chain of Being*. 12th ed. Cambridge: Harvard University Press, 1936.

Macdonald, Allan John. *Berengar and the Reform of Sacramental Doctrine*. London: Longmans, Green and Co., 1930.

Maclagan, E.R.D., and A.G.B. Russell, eds. *The Prophetic Books of William Blake*. London: A.H. Bullen, 1904.

Malet, Andre. *The Thought of Rudolf Bultmann*. Trans. Richard Strachan. Shannon: Irish UP, 1969.

Malkan, Jeffrey. "Blake's 'The Mental Traveller' as Proto-Novel." Unpublished paper, 1978.

McFarland, Thomas. *Coleridge and the Pantheist Tradition*. Oxford: Clarendon, 1969.

McGann, Jerome J. *The Romantic Ideology: A Critical Investigation*. Chicago: Univ. of Chicago, 1983.

McKusick, James. *Coleridge's Philosophy of Language*. Yale, 1986.

Mellor, Anne Kostelantz. *Blake's Human Form Divine*. Berkeley: University of California Press, 1974.

———. "*Frankenstein* and the Sublime." In Behrendt, *Approaches*. 99–104.

———, ed. *Romanticism and Feminism*. Bloomington: Indiana UP, 1988.

———. *Romanticism and Gender*. NY: Routledge, 1993.

Miller, Dan. "Blake and the Deconstructive Interlude." In Miller, *Critical Paths*. 139–67.

Miller, Dan, Mark Bracher, and Donald Ault, eds. *Critical Paths: Blake and the Argument of Method*. Durham: Duke University Press, 1987.

Miller, J. Hillis. "The Critic as Host." In Bloom, *Deconstruction*. 217–53.

Miller, Madeleine S., and J. Lane Miller, eds. *Harper's Bible Dictionary*. NY: Harper and Brothers, 1956.

Milton, John. *Paradise Lost*. In *Complete Poems and Major Prose*. Ed. Merritt Y. Hughes. New York: Odyssey Press, 1957.

Miner, Paul. "Visionary Astronomy." *Bulletin of Research in the Humanities* 84 (1981): 305–36.

———. "Visions in the Darksom Air: Aspects of Blake's Biblical Symbolism." *William Blake*. Ed. Rosenfeld. 256–92.

Mitchell, W.J.T. *Blake's Composite Art: A Study of the Illuminated Poetry*. Princeton UP, 1978.

———. "Visible Language: Blake's Wond'rous Art of Writing." In Eaves and Fischer. 46–95.

Miyoshi, Masao. *The Divided Self: A Perspective on the Literature of the Victorians*. NY: NYUP, 1989.

Moore, Thomas. *Care of the Soul: A Guide for Cultivating Depth and Sacredness in Everyday Life*. HarperCollins, 1992.

Morton, A.L. *The Everlasting Gospel: A Study in the Sources of William Blake*. London: Lawrence and Wishart, 1958.

Nathan, Norman. *Prince William Blake: The Philosophical Conceptions of William Blake.* The Hague: Mouton, 1975.

The New English Bible, with the Apocrypha. London: Oxford UP, 1970.

The New Smith's Bible Dictionary. Ed. William Smith, revised by Reuel G. Lemmons. New York: Doubleday, 1966.

Newman, Beth. "The Frame Structure of *Frankenstein.*" *English Literary History* 53 (1986): 141–63.

Nurmi, Martin K. "Joy, Love, and Innocence in Blake's 'The Mental Traveller'." *Studies in Romanticism* 3 (1964) 109–17.

O'Neill, Michael. "Shelley's *The Triumph of Life*: Questioning and Imagining." In Watson, *Infinite.* 161–80.

Ostriker, Alicia. *Vision and Verse in William Blake.* Madison: Univ. of Wisconsin Press, 1965.

———, ed. *William Blake: The Complete Poems.* NY: Penguin, 1977.

Otto, Peter. *Constructive Vision and Visionary Deconstruction: Los, Eternity, and the Productions of Time in the Later Poetry of William Blake.* Oxford, 1991.

Pagels, Elaine. "The Orthodox Against the Gnostics: Confrontation and Interiority in Early Christianity." In Berger, *The Other Side.*

Pagliaro, Harold. *Selfhood and Redemption in Blake's* Songs. University Park: The Pennsylvania State UP, 1987.

Paley, Morton D. *The Continuing City.* Oxford: Clarendon, 1983.

———. "The Female Babe and 'The Mental Traveller'." *Studies in Romanticism* 1 (1962): 97–104.

Parrish, Stephen Maxfield, ed. *Coleridge's Dejection: The Earliest Manuscripts and the Earliest Printings.* Ithaca: Cornell UP, 1988.

Parvey, Constance F. "The Theology and Leadership of Women in the New Testament." In Ruether, *Religion and Sexism.* 117–49.

Pechey, Graham. "*The Marriage of Heaven and Hell*: A Text and Its Conjuncture." *Oxford Literary Review* 3 (1979): 52–76.

Piquet, Francois. "*Le Voyageur Spirituel* et le Cercle de Destinée Blakien." Bertrandias, Bernadette, Christian La Cassagnere, and Jean Raimond, eds. *Aspects du Romantisme Anglais. Mélanges Offerts à Jacques Blondel.* Université de Clermont, 1980.

Poovey, Mary. *The Proper Lady and the Woman Writer: Ideology as Style in the Works of Mary Wollstonecraft, Mary Shelley, and Jane Austen.* Chicago: Univ. of Chicago Press, 1984.

Pope, Alexander. *An Essay on Man.* Ed. Maynard Mack. Twickenham Edition. 1950, rpt. New Haven: Yale University Press, 1964.

Popper, Karl R. "What Is Dialectic?" *Conjectures and Refutations.* London, 1963. 312–35.

Poulet, Georges. *The Metamorphoses of the Circle.* Trans. Carley Dawson and Elliott Coleman. Baltimore: Johns Hopkins University Press, 1966.

Prickett, Stephen. "*Biographia Literaria*: Chapter Thirteen." In Coleman and Otto. 3–23.

———. *Romanticism and Religion: The Tradition of Coleridge and Wordsworth in the Victorian Church.* Cambridge: Cambridge UP, 1976.

———. *Words and the Word: Language, Poetics and Biblical Interpretation.* Cambridge: Cambridge UP, 1986.

Pulos, C.E. *The Deep Truth: A Study of Shelley's Scepticism.* Univ. of Nebraska Press, 1954.

Punter, David. "Blake and the Shapes of London." *Criticism* 23 (1981): 1–23.

Raine, Kathleen. *Blake and Tradition.* 2 vols. Princeton: Princeton UP, 1968.

Reiman, Donald. *Shelley's "The Triumph of Life": A Critical Study.* Urbana: U. of Illinois Press, 1965.

Rescher, Nicholas, and Robert Brandom. *The Logic of Inconsistency: A Study in Non-Standard Possible-World Semantics and Ontology*. Oxford: Basil Blackwell, 1980.

Riccati, Carlo. *"Processio" et "Explicatio": La doctrine de la création chez Jean Scot et Nicolas de Cues*. Naples: Bibliopolis, 1983.

Richardson, Alan. "From *Emile* to *Frankenstein*: The Education of Monsters." *European Romantic Review* 1 (Winter 1991): 147–62.

Ricoeur, Paul. *Essays on Biblical Interpretation*. Philadelphia: Fortress Press, 1980.

Riede, David G. "The Symbolism of the Loins in Blake's *Jerusalem*." *Studies in English Literature* 21 (1981): 547–61.

Rose, Edward J. "Blake's Human Root: Symbol, Myth, and Design." *Studies in English Literature* 20 (1980): 575–86.

———. "Circumcision Symbolism in Blake's *Jerusalem*." *Studies in Romanticism* 8 (1968): 16–25.

———. "The Meaning of Los." *Blake Newsletter* 3 (1967): 10–11.

———. "The Structure of Blake's *Jerusalem*." *Bucknell Review* 11 (1962): 35–54.

Rosenfeld, Alvin H., ed. *William Blake: Essays for S. Foster Damon*. Providence: Brown UP, 1969.

Roston, Murray. *Prophet and Poet: The Bible and the Growth of Romanticism*. Evanston: Northwestern UP, 1965.

Rubin, Merle R. "Shelley's Skepticism: A Detachment Beyond Despair." *Philological Quarterly* 59 (1980): 353–73.

Rudd, Margaret. *Organiz'd Innocence: The Story of Blake's Prophetic Books*. London: Routledge and Kegan Paul, 1956.

Rudens, S.P. "The God of William Blake." *The Reflex* 4 (1929): 36–41.

Ruether, Rosemary Radford. "Misogynism and Virginal Feminism in the Fathers of the Church." In Ruether, *Religion and Sexism*. 150–83.

———, ed. *Religion and Sexism: Images of Woman in the Jewish and Christian Traditions*. NY: Simon and Schuster, 1974.

Salvat, Michel. "Les Eléments Chrétiens dans la Pensée de William Blake." *La Vie des Peuples* 54 (1924): 330–59.

Sandmel, Samuel. *The Hebrew Scriptures: An Introduction to Their Literature and Religious Ideas*. NY: Knopf, 1963.

Schneidau, Herbert N. *Sacred Discontent: The Bible and Western Tradition*. Baton Rouge: Louisiana State UP, 1976.

Schug, Charles. "The Romantic Form of Mary Shelley's *Frankenstein*." *Studies in English Literature* 17 (1977): 607–19.

Shelley, Mary. *Frankenstein*. Ed. M.K. Joseph. NY: Oxford UP, 1969.

Shelley, Percy Bysshe. *Shelley's Poetry and Prose*, ed. Donald H. Reiman and Sharon B. Powers. NY: Norton, 1977.

Sigmund, Paul E. *Nicholas of Cusa and Medieval Political Thought*. Cambridge: Harvard UP, 1963.

Singer, Dorothea Waley. *Giordano Bruno: His Life and Thought. With Annotated Translation of His Work* Of the Infinite Universe and Worlds. NY: Henry Schuman, 1950.

Sloss, D.J., and J.P.R. Wallis, eds. *The Prophetic Writings of William Blake*. Oxford: Clarendon, 1926.

Smith, Cyprian. *The Way of Paradox: Spiritual Life as Taught by Meister Eckhart*. London: Darton, Longman and Todd, 1987.

Smith, D. Moody, Jr. "The Use of the Old Testament in the New." In Efird, *The Use of the Old Testament*.

Smith, Mark Trevor. "Striving with Blake's Systems." In Erdman, *Blake and His Bibles*. 157–78.

Snyder, Alice D. *The Critical Principle of the Reconciliation of Opposites as Employed by Coleridge*. 1918; rpt. Ann Arbor: Folcroft, 1970.

Southern, Richard William. "Lanfranc of Bec and Berengar of Tours." In Hunt, R.W., W.A. Pantin, and R.W. Southern, eds., *Studies in Medieval History: Presented to Frederick Maurice Powicke*. 1948; rpt. Oxford: Clarendon Press, 1969. Pp. 27–48.

Spitzer, Leo. *Classical and Christian Ideas of World Harmony.* Ed. Anna Granville Hatcher. Baltimore: Johns Hopkins University Press, 1963.

Summerfield, H. "Blake and the Names Divine." *Blake: An Illustrated Quarterly* 57 (Summer 1981): 14–22.

Sutherland, John H. "Blake's 'Mental Traveller'." *English Literary History* 22 (1955): 136–47.

Swingle, L.J. "Frankenstein's Monster and Its Romantic Relatives: Problems of Knowledge in English Romanticism." *Texas Studies in Literature and Language* 14 (1973): 51–65.

Tachau, Katherine H. *Vision and Certitude in the Age of Ockham: Optics, Epistemology and the Foundations of Semantics, 1250–1345.* Leiden: E.J. Brill, 1988.

Tannenbaum, Leslie. *Biblical Tradition in Blake's Early Prophecies: The Great Code of Art.* Princeton: Princeton UP, 1982.

Tebeaux, Elizabeth. "Scepticism in Pope's 'Essay on Man.'" *College Literature* 10 (1983): 158–71.

Thorslev, Peter L. *Romantic Contraries: Freedom versus Destiny.* New Haven: Yale UP, 1984.

———. "Some Dangers of Dialectic Thinking, with Illustrations from Blake and His Critics." *Romantic and Victorian: Studies in Memory of William H. Marshall.* Ed. Elledge, W. Paul, and Richard L. Hoffman. Rutherford: Fairleigh Dickinson UP, 1971. 43–74.

Tolley, Michael John. "William Blake's Use of the Bible." Diss. University of London, 1974.

Tropp, Martin. *Mary Shelley's Monster: The Story of Frankenstein.* Boston: Houghton Mifflin, 1976.

Tuve, Rosemund. "A Mediaeval Commonplace in Spenser's Cosmology." *Studies in Philology* 30, no. 2 (1933): 133–47.

Twitchell, James B. "'The Mental Traveller,' Infinity and the 'Arlington Court Picture.'" *Criticism* 17 (1975): 1–14.

Vansteenberghe, Edmond. *Autour de la Docte Ignorance: Une Controverse sur la Théologie Mystique au XVe Siècle.* Munster en Westphalie: Imprimerie Aschendorff, 1914.

———. *Le Cardinal Nicolas de Cues.* 1920; rpt. Frankfurt: Minerva GMBH, 1963.

Von Franz, M.-L. "The Process of Individuation." In Jung, Carl G., ed. *Man and His Symbols.* 1964; rpt. NY: Dell, 1970.

Wagenknecht, David. "Afterword." In Miller, *Critical Paths.* 310–28.

———. *Blake's Night: William Blake and the Idea of Pastoral.* Cambridge: Belknap Press of Harvard UP, 1973.

Walling, William. "*Frankenstein* in the Context of English Romanticism." In Behrendt, *Approaches.* 105–11.

Warnock, Mary. *Imagination.* Berkeley: Univ. of California Press, 1976.

Wasserman, Earl. *Shelley's* Prometheus Unbound. Baltimore: Johns Hopkins University Press, 1965.

———. *The Subtler Language.* 1959; rpt. Baltimore: Johns Hopkins University Press, 1964.

Watson, J.R., ed. *An Infinite Complexity: Essays in Romanticism.* Edinburgh UP, 1983.

Watts, Pauline Moffitt. *Nicolaus Cusanus: A Fifteenth-Century Vision of Man.* Leiden: E.J. Brill, 1982.

Wheeler, Kathleen M. *Sources, Processes and Methods in Coleridge's* Biographia Literaria. Cambridge: Cambridge UP, 1980.

White, Douglas H., and Thomas P. Tierney. "*An Essay on Man* and the Tradition of Satires on Mankind." *Modern Philology* 85 (1987): 27–41.

Wicksteed, Joseph. *William Blake's Jerusalem: A Commentary.* London: Trianon, n.d.

Wiley, Margaret L. *The Subtle Knot: Creative Scepticism in Seventeenth-Century England.* Cambridge, Harvard UP, 1952.

Wittreich, Joseph Anthony, Jr. *Angel of Apocalypse: Blake's Idea of Milton.* Madison: University of Wisconsin Press, 1975.

———, ed. *Nineteenth-Century Accounts of William Blake.* Gainesville, FL: Scholars' Facsimiles and Reprints, 1970.

———. *Visionary Poetics: Milton's Tradition and His Legacy.* San Marino, CA: Huntington Library, 1979.

Wolfson, Susan J. "Feminist Inquiry and *Frankenstein.*" In Behrendt, *Approaches.* 50–59.

Woodhouse, A.S.P. *The Poet and His Faith: Religion and Poetry in England from Spenser to Eliot and Auden.* Chicago: University of Chicago Press, 1965.

Woodman, Ross. "The Death and Resurrection of Milton According to the Gospel of Blake." *English Studies in Canada* 3 (1977): 416–32.

Wordsworth, William. *William Wordsworth.* The Oxford Authors. Ed. Stephen Gill. Oxford UP, 1984.

Yeats, W.B. *The Poems. A New Edition.* Ed Richard J. Finneran. NY: Macmillan, 1983.

———. *A Vision.* London: Macmillan, 1937.

Zoellner, Robert H. "Poetic Cosmology in Pope's *An Essay on Man.*" *College English* 19 (1958): 157–62.

Index

The Locust Hill Literary Studies Series

1. *Blake and His Bibles*. Edited by David V. Erdman. ISBN 0-933951-29-9. LC 89-14052.
2. *Faulkner, Sut, and Other Southerners*. M. Thomas Inge. ISBN 0-933951-31-0. LC 91-40016.
3. *Essays of a Book Collector: Reminiscences on Some Old Books and Their Authors*. Claude A. Prance. ISBN 0-933951-30-2. LC 89-12734.
4. *Vision and Revisions: Essays on Faulkner*. John E. Bassett. ISBN 0-933951-32-9. LC 89-14046.
5. *A Rose by Another Name: A Survey of Literary Flora from Shakespeare to Eco*. Robert F. Fleissner. ISBN 0-933951-33-7. LC 89-12804.

6 *Byron's Poetic Technique*. David V. Erdman. ISBN 0-933951-39-6.

7. *Blake's Milton Designs: The Dynamics of Meaning*. J.M.Q. Davies. ISBN 0-933951-40-X. LC 92–32678.
8. *The Slaughter-House of Mammon: An Anthology of Victorian Social Protest Literature*. Edited by Sharon A. Winn and Lynn M. Alexander. ISBN 0-933951-41-8. LC 92-7269.
9. *"A Heart of Ideality in My Realism" and Other Essays on Howells and Twain*. John E. Bassett. ISBN 0-933951-36-1. LC 90-46908.
10. *Imagining Romanticism: Essays on English and Australian Romanticisms*. Edited by Deirdre Coleman and Peter Otto. ISBN 0-933951-42-6. LC 91-36509.
11. *Learning the Trade: Essays on W.B. Yeats and Contemporary Poetry*. Edited by Deborah Fleming. ISBN 0-933951-43-4. LC 92–39290.
12. *"All Nature is but Art": The Coincidence of Opposites in English Romantic Literature*. Mark Trevor Smith. ISBN 0-933951-44-2. LC 93–27166.
13. *Essays on Henry David Thoreau: Rhetoric, Style, and Audience*. Richard Dillman. ISBN 0-933951-50-7. LC 92–39960.